HARNESSING THE AIRPLANE

Harnessing the
AIRPLANE

American and British Cavalry Reponses
to a New Technology, 1903–1939

LORI A. HENNING

UNIVERSITY OF OKLAHOMA PRESS : NORMAN

This book is published with the generous assistance of the McCasland Foundation, Duncan, Oklahoma.

LIBRARY OF CONGRESS CATALOGING-IN-PUBLICATION DATA

Names: Henning, Lori A., author.

Title: Harnessing the airplane : American and British cavalry responses to a new technology, 1903–1939 / Lori A. Henning.

Description: First edition. | Norman, OK : University of Oklahoma Press, [2019] | "This book examines the ramifications of technological innovation and its role in the fraught relationship that developed between traditional ground units and emerging air forces in the American and British armies during the early twentieth-century."—Provided by publisher. | Includes bibliographical references and index.

Identifiers: LCCN 2018027410 | ISBN 978-0-8061-6184-6 (hardcover : alk. paper)

Subjects: LCSH: United States. Army. Cavalry—Drill and tactics. | Great Britain. Army—Cavalry—Drill and tactics. | United States. Army. Cavalry—History— 20th century. | Great Britain. Army—Cavalry—History—20th century. | United States. Army. Air Corps—History—20th century. | Great Britain. Royal Air Force—History—20th century.

Classification: LCC UE160 .H46 2019 | DDC 357/.184—dc23

LC record available at https://lccn.loc.gov/2018027410

The paper in this book meets the guidelines for permanence and durability of the Committee on Production Guidelines for Book Longevity of the Council on Library Resources, Inc. ∞

To my parents
Thank you for your love and support,
which made all of my successes possible,
including this book.

CONTENTS

ILLUSTRATIONS

ACKNOWLEDGMENTS

This work would not have been completed without the support and assistance of many organizations and individuals. I extend my gratitude to Mr. and Mrs. Roger Jenswold and the Smithsonian Air and Space Museum (NASM) for their financial contributions supporting my research. My appreciation also goes to the Smithsonian Air and Space writing group, which showed me that even senior scholars struggle with the writing process. Thanks also go to the many librarians, archivists, and employees at the Air Force Historical Research Agency, the British National Archives, the U.S. Army Military History Institute, the U.S. National Archives in College Park and my NASM sponsor, Alex Spencer, who helped me with my research.

I also thank the many supportive faculty members at Texas A&M University who advised and assisted me along the way, including Chester Dunning, Brian McAllister Linn, Adam Seipp, David Vaught, and most especially Jonathan Coopersmith. As my advisor, Coopersmith, or the "mighty JC" as I referred to him unbeknownst to him, exemplified the kind of advisor I endeavor to become, because he was concerned with his students' entire well-being. He always asked about my physical and mental health before asking about the progress of my dissertation, perhaps prompted by my heavy breathing after walking up the three flights of steps that separated our offices. Although I did not think about it at the time, his encouragement to bike to school (quite an undertaking in the summer in Texas), sleep more than a few hours a night, and remember to eat had vast benefits beyond the physical.

I would be remiss to not also express my gratitude to my friends, family, and colleagues who also supported me throughout my dissertation and manuscript writing process, especially James Anderman; my parents, Louis and Linda Henning; Roger Horky, and David Lucsko.

HARNESSING THE AIRPLANE

Introduction

The first time I saw a Segway was around 2006 on the campus of Texas A&M University. I could not resist the temptation to stare and smile as the sixty-ish-year-old man wearing his bicycle helmet zoomed around campus. Each time I saw him in the months that followed, I watched the students watching him. They stared, pointed, and smiled or laughed as this middle-aged man rolled around campus. Inventor Dean Kamen's prediction that the Segway would "be to the car what the car was to the horse and buggy" seemed increasingly unrealistic at the time as I observed the students' reactions.[1] Now more than ten years later, skeptics of the potential transformative nature of the Segway appear to have been correct, but what if future events prove them wrong? How will scholars' assessments of the character of these early doubters' change?

When a technology does succeed, few concentrate on the length of time it may have taken for the innovation to supplant what came before. Despite the seeming failure of his 2001 prediction of Segway's transformation of the world, Kamen was still confident that the conclusion of the story had not been written in 2006. In an article in *Time,* he reminded the reader that airplanes were in a niche market after they were first introduced, and it took time for their revolutionary effect to be realized.[2] In the case of airplanes, aviation historians tend to gloss over the early days of tactical aviation and jump to World War I aerial dogfights and the first bombs dropped on targets behind the front lines. Military airplanes did not have a particular role or function when they were first built. It took time, measured in years, to determine their use and design and to construct and modify the technology for various uses. Historians who ignore the sometimes-long duration of debates over the use of a new innovation prevent a full understanding of how technological incorporation occurs. Even if the design of a specific technology does not change, its possible uses can. Sometimes it takes years to decide the appropriate use for a technology.

The label "technological resisters" is often derogatory, intended to portray such people as the villains in the grand narrative of technological progress. Perhaps exemplifying the cliché that victors write the history, scholars usually write about successful technologies. Most, including historians of technology, do not satisfactorily address the large number of technological ideas that fail.[3] The lack of this scholarship is more glaring for the first three decades of the twentieth century because of the innovativeness of this period. The increase in the number of accepted transformative technologies probably occurred not because the quality of inventions increased but because the number of technologies being developed increased.[4] There are, however, many obstacles to studying unsuccessful technologies, including embarrassment, lack of sources, and forgetfulness—both deliberate and otherwise. The lack of scholarship on failed technology complicates the ability to appreciate the character and significance of technological successes and how they are examined. For example, if most histories of technology are about successful technologies, they may produce the false impression that technologies are bound to succeed.[5] If the idea that new technologies are bound to succeed is accepted, even unconsciously, scholars may tend to treat those who resisted a new technology more harshly or critically than the historical situation warrants. Because most innovations do not succeed, those who resisted a new technology may prove to have been more sensible than those who advocated blind adoption in the mistaken belief that it must automatically be better. Of course, the only way resisters can be accused of being backward is after the fact.

This work tells the story of cavalrymen and their supporters, a group of these "resisters," who were transformed over decades into accepters of a transformative technology. This project is partly a response to historian Adrian Randall's call to appreciate that "much fundamental change is painful and that which destroys old ways of life more painful still" and to encourage scholars to treat those "who resisted change with more humility and with more sympathy."[6] This project examines a group that critically examined airplanes and subsequently suffered criticism from those who had the benefit of knowing the end of the story.

Lacking the perfect hindsight of historical perspective, those confronted with a new technology must make their decisions based on insufficient information, which is not easy. The fact that innovators and those controlling the technology idealize innovations, particularly new weapon systems' value and capabilities, complicates proper assessment of a technology's capabilities.[7] The problem is that few contemporaries of a technology can accurately predict which technologies will be transformative and which will fail. Historian Edward Katzenbach provided one of the best descriptions of this difficulty when he wrote that the "crystal ball" that gave people the ability to predict what future war would be like "has been shattered by technology."[8] The airplane was among the revolutionary technologies that destroyed the ability to foresee the future.

Historians have produced excellent studies focused on early enthusiasts who unquestioningly embraced the airplane in its infancy. Few, however, provide a thorough examination of the ideas and attitudes that prevailed among those realists who questioned the overly enthusiastic and exaggerated predictions of aviation proponents. By analyzing the development of aviation in relation to the cavalry, this work examines one group within the larger population of the "minority of commentators," mentioned by Joseph Corn in *The Winged Gospel*, which examines Americans' responses to the advent of aviation. Cavalrymen were among those who raised what Corn called a "note of caution and skepticism amidst the din of unrestrained prophecy greeting the airplane." Refusing to accept the predictions of aviation supporters on faith alone did *not* make cavalrymen "anti-airplane or even negative toward the invention's impact."[9] Cavalrymen enumerated the innovation's limitations, dismissing the most outrageous claims, but they also envisioned how cooperation with and utilization of airplanes would benefit their branch. Therefore, the pejorative label "naysayers" frequently employed against those critical of a new technology is not accurate. A better moniker for this group is "cautious technological examiners" because many cavalrymen welcomed the introduction of aviation into the military once airplanes' true capabilities were discerned.

Unlike previous studies conducted by army or cavalry scholars where the airplane is simply one technology in the long litany of technologies the cavalry confronted, this work explores the unique challenges posed by airplanes.[10] Unlike the other technologies—including tractors, trucks, automobiles, motorcycles, armored scout cars, and tanks—airplanes are not a ground-based threat. Although it is a distinctive technology, the airplane may prove to be the model of technological change that may provide a better understanding of cavalrymen's response to tanks and armored cars than similar examinations of trucks, motorcars, and bicycles. Similar tactics to those used in response to airplanes were utilized in debates on tanks.

This work examines how American and British cavalrymen responded to the unique threat and opportunity posed by airplanes from 1903 through 1939.[11] They perceived aviation as both a threat and potential opportunity, their opinions shifting over time as aviation technology improved and their understanding of their branch's roles and missions changed. Other works have described how various cavalry organizations responded to another new technology, the tank. What is often forgotten is that the horse had two potential successors—the tank and airplane—and that the organizations championing strategic bombing also created a doctrine of tactical observation that connected aviation directly to the cavalry's missions. This work will reestablish the link between cavalry and aviation in the first four decades of the twentieth century. It reveals a horse cavalry that worked with technological change, embraced the airplane in some cases and experimented in others, established doctrine and applications for joint operations, and even tried to develop its own air contingent of autogiros. The cavalry, like many old and established institutions, had some reactionaries who preferred the branch's traditions and rejected new ideas, but most cavalrymen were pragmatic. They understood that technological change was inevitable and wanted to manage the process for the benefit of their branch.

Aviation historians overlook the link between aircraft and cavalry, focusing almost entirely on the development of air power doctrine, which has dominated scholarship on the 1920s and 1930s. They too overlook that the earliest military aircraft were employed in reconnaissance and observation roles, concentrating instead on advancements in technology, air-to-air combat, and strategic bombing. Rarely is any attention paid to the overlap of airplanes and cavalry for tactical observation missions and the support of ground troops.[12] Instead, they emphasize air operations independent of surface units; therefore, any mention of air power in the service of land warfare is neglected.[13] As a result, the connections between aviation and the cavalry have been slighted. Some aviation histories briefly mention the connection between the airplane's early use as a reconnaissance platform and cavalry's role as the eyes and ears of the army, but they quickly turn to technical discussions of aerial photography, cameras, or mapmaking with little mention of how these developments affected the cavalry.

It is not yet clear why a study of this kind has not been done. It may be that the tank has overshadowed and continues to eclipse coverage of the airplane because the tank as a land-bound vehicle was the obvious successor to the horse. Furthermore, it may be that many cavalry officers at the time did not clearly differentiate between aviation and mechanization as different forms of modernization, which led historians to the same conclusion. It may also be a result of the bomber mafia dominating aviation doctrine for most of the first half of the twentieth century.

Whatever the reason, this work's loosely chronological chapters analyze the connection between the cavalry and aviation during the early years of heavier-than-air flight. The chapters demonstrate that cavalrymen and their supporters pragmatically responded to the introduction of airplanes. Cavalrymen critically examined the utility of flying machines by assessing their capabilities in war and peace.

Beginning with a brief history of how the American and British cavalries modernized to meet the demands of modern warfare prior to their nation's acquisitions of military heavier-than-air aircraft in 1909 and 1908, respectively, this book traces the responses of American and British cavalrymen to the rapid changes in aviation technology and doctrine from 1903 through 1939. It compares and contrasts the American and British cavalrymen's evolving responses to military airplanes as aviation technology improved and became capable of assisting in missions previously accomplished solely by the cavalry. Cavalrymen in both nations were cautious in assessing the limited but rapidly evolving capabilities of military airplanes. At first, they saw airplanes as having far too many limitations and drawbacks to fulfill the predictions made by aviation supporters, but nonetheless they discussed them, experimented with them in maneuvers, and found them at times capable of aiding the mounted branch.

The cavalry's records show a cavalry not opposed to the airplane, aviation, aviators, or mechanization advocates on general principles, but a group defending itself against the popular belief that it was obsolete. Discussions of the limitations of new technologies had less to do with a distrust or hatred of technology and more with debunking the theories of overly optimistic supporters of modern weapons, elicited in part by national movements for economy and modernization. The cavalry made efforts to incorporate aviation technology well into the 1930s (and the early 1940s in the case of the United States). This desire was clear in the cavalry's testing of autogiros, a type of experimental aircraft. Abandoned by the Army Air Forces, which were focused on strategic bombing and other independent operations, not surface unit support, the cavalry was still attempting to establish its own air section when World War II commenced. The American and British cavalries were not threatened by aviation; in fact, they endeavored to embrace it, experimenting with new air machines to maintain their proficiency and, by extension, their existence.

In addition to exploring the actions of cavalrymen, this work also compares and contrasts what aircraft and horse cavalry were actually able to accomplish with popular contemporary assertions about their capabilities. Although cavalrymen and their supporters' evaluations were frequently more accurate than those of air advocates, the cavalrymen of the United States and Great Britain were increasingly characterized as conservative diehards unwilling to accept anything new. In the end, cavalrymen were becoming desperate, relying less on logic and more on sentiment. In addition to reasoned arguments, some cavalrymen's justifications for their service included poems decrying the loss of the horse to the unfeeling machine, which may be one reason why the modern perception of cavalrymen is so negative.

Despite many similarities between British and American military visions of aviation, each chapter illuminates noteworthy differences between the two that produced divergent results. The United States and Great Britain faced two different sets of strategic challenges throughout the first half of the twentieth century that directly affected how the cavalries of each country responded to the introduction of aviation. Their histories, experiences, and culture directly affected the way militaries eventually eliminated their horse cavalries. The United States was behind its ocean border with only minor intermittent problems with Mexico and small commitments outside of its continental location. Britain, which had relied almost entirely on its navy for the security of its home island, was forced to rethink its defensive strategy when the airplane was introduced. It also had to concern itself with policing its extensive empire. Each country had different demands that called for different budgeting and concentration.

The conclusion of this work ties this historical experience with general insights into how organizations respond to novel technologies that threaten to alter or eliminate them. The theoretical debates about the relative merits of man and machine were less important than the practical day-to-day challenges of limited budgets, uncertainty about enemy plans, and technological limitations.

CHAPTER ONE

State of Affairs

The United States and Great Britain took slightly different paths in their response to aviation's effect on the cavalry. Their discrete routes stemmed from their dissimilar perspectives on the cavalry's roles in modern warfare, their unique experience with aviation, and especially their cavalry's historical organization. British and American cavalrymen's responses to aviation stemmed directly from what each nation believed was the cavalry's appropriate role in modern warfare. This belief was the product of each nation's history and recent efforts at modernizing and professionalizing its branches. In keeping with the late-nineteenth-century trend of professionalization, the British and American cavalries founded military service schools and professional societies, studied military history and recent conflicts, and published journals. The two institutions approached these tasks differently, however, based on the differences between their nations and mounted forces, which included traditions grounded in the history of their branches, organization of their mounted forces, wartime experiences, and public opinion. Substantial differences existed within the American and British cavalry communities regarding mounted versus dismounted tactics; the use of lances, sabers, and firearms; and the relative importance of the charge, the raid, and reconnaissance. Early experiments and experiences with airplanes also differed. These dissimilarities are key to understanding why each nation's cavalrymen responded the way they did to aviation.

United States Cavalry before 1903

Unlike many European mounted units, the United States Cavalry at the turn of the twentieth century had a short and noncontinuous history. It had also developed a unique "American way" of using cavalry that was different from the traditional view of cavalry in Europe. The European concept of the cavalryman as a mounted warrior charging the enemy with sword drawn contrasted with American cavalry use. American cavalrymen usually fought on foot and rode horses primarily to increase their mobility.[1] This was true in the American Revolutionary War, during disturbances on the western frontier and local troubles, and in the Mexican-American War. The 1st Regiment of Dragoons, mounted for speed and able to fight mounted or dismounted, was a more versatile force than the Mounted Rangers it replaced in March 1833, who could fight only when dismounted.[2] The Dragoons, however, largely retained the character of mounted infantrymen, spending more time fighting dismounted and being seldom used as a shock weapon.[3] In the Mexican-American War (1846–48), as in the Revolutionary War, reconnaissance and pursuit missions dominated. There were few cavalry charges despite the branch's participation in all of the war's major campaigns.[4]

This American way of using cavalry more clearly emerged during the American Civil War and in postwar conflicts with Native Americans.[5] Brigadier General John Buford's use of the Union's First Division Cavalry Corps on the first day of the Battle of Gettysburg in 1863 exemplified this innovative use of mounted soldiers. Buford dismounted his cavalry to defend the favorable ground outside Gettysburg against a Confederate infantry division until the Federal infantry arrived.[6] In the war, the Union cavalry, as well as the Confederate cavalry, served as a sort of jack-of-all-trades, filling roles requiring both mobility and staying power. Cavalry proved invaluable in reconnaissance and counterreconnaissance when under the capable leadership of officers such as Buford and Philip Sheridan.[7] American cavalrymen were not just shock troops, pickets, and scouts but, as one historian accurately summarized, "highly mobile gunmen who could use their horses to deny strategic positions to the enemy and hold them with rapid-fire repeating carbines until infantry support arrived."[8] Following the war, American cavalrymen pursued Native Americans on horseback, with chases frequently ending in a dismounted fight.[9] In contrast to European contemporaries, a pragmatic doctrine evolved in the United States based on recent conditions, not historical examples of massed cavalry charges. American mounted soldiers were not trained exclusively to fight mounted, nor did they. The cavalry was rarely concentrated in a large force or put into situations that made charging desirable.

Serious attempts to formalize this method of cavalry employment in written doctrine emerged later, during the 1880s as part of the U.S. military modernization program. Inspired by the desire to survive a new political and economic age, the U.S. military modernized its training, organization, and officers.[10] This transformation mirrored the professionalization occurring in other sectors of society. Professionalization included the formation of several schools to produce well-trained and knowledgeable officers.[11]

In 1881, the War Department founded the School of Application for Infantry and Cavalry at Fort Leavenworth, Kansas, to train junior officers in various professional military subjects, particularly small-unit tactics. The curriculum focused on the military arts of tactics, strategy, logistics, and military history.[12] The school disseminated the precepts of these branches throughout the army by assigning one lieutenant from each cavalry and infantry regiment every two years.[13] In 1892, the War Department opened a similar school for cavalry and light artillery at Fort Riley, Kansas.[14] The army opened more than thirty additional schools for various specialties by the end of the Great War.[15]

In their early years, these schools developed useful curricula that included instruction in the varied use of the cavalry. The "Programm [sic] of the Course of Cavalry" at the Infantry and Cavalry School consisted of recitations, drills, problems, and field exercises in tactics, field service, equitation, and hippology (the study of horses).[16] The training for the cavalry was almost identical to the "Infantry Programme" and stressed both mounted and dismounted action, corresponding to the cavalry's experience on the western frontier and during the Civil War.[17]

Due to the writings of a few military theorists and officers, the military schools over the next twenty years evolved into institutions where officers could learn the special skills of their branch of service as well as the attitudes and principles of higher command.[18] Leavenworth instructor and cavalryman Arthur L. Wagner, dubbed the "American Pioneer in the Cause of Military Education" by Eben Swift, a similarly minded colleague, was one of the many champions of reform and helped formalize cavalry policy.[19] Wagner's works, written at Fort Leavenworth during his tenure there (1888–1904), included the often-reprinted 1895 *Organization and Tactics* and the 1893 *Service of Security and Information*, which detailed the tactical employment of infantry, cavalry, and artillery in both offensive and defensive situations.[20] Other works utilized at the military postgraduate schools were combined with Wagner's works to create a core of knowledge.[21]

Although Wagner's works emphasized the infantry, he noted the skill of cavalry in reconnaissance, raiding, and pursuit.[22] He urged flexibility and promoted the jack-of-all-trades use of the cavalry. Wagner helped to condense much of the wisdom of cavalry use developed during the Civil

War. His writings concentrated on the cavalry's nonbattlefield duties, emphasizing the mounted branch's responsibilities to the army on the move.[23] These duties included the cavalry becoming both a screen and the main reconnaissance force of the army, keeping commanders informed of the enemy's positions and denying similar information to the enemy.[24] Wagner argued in an abridgement of *The Service of Security and Information* that the cavalry could conduct reconnaissance better than infantry because the greater mobility of the cavalry meant that the "reconnoitering duty can be performed more efficiently and more easily by cavalry than by infantry" and a strong cavalry screen increased the army's security.[25]

In recommending the cavalry's use as an advanced guard, Wagner noted a marked difference between the training and roles of the American cavalry and those of its European counterparts. The distance between the main army force and advanced guard of cavalry in the American army could be reduced compared to other armies because the cavalry could "make effective use of dismounted fire-action, [because it] has greater resisting power than European cavalry, and it is not limited, as the latter seems generally to be, to a charge to the front or a flight to the rear."[26] Further, he explained,

> all European authorities recommend the use of cavalry as reconnoiterers, but prescribe that the support should consist in part of infantry to supply the necessary resisting power. In our service this is not in general necessary, as our cavalry has enough resisting power to carry out the delaying action of the support; and nothing but the lack of sufficient cavalry should necessitate the adoption of a composite support. Indeed it is, in most cases, a great mistake so to combine cavalry with infantry as to tie the former down to the pace of the latter.[27]

Wagner argued the cavalry should rely mainly on dismounted fire action, which was what "good cavalry of the American type" did.[28] Even while continuing to maintain that shock action was the cavalry's primary role, to be used in every suitable situation, Wagner suggested the use of the carbine when the cavalry was on the defensive against a well-concentrated opponent.[29] He stated that on the battlefield and in defending the rear of the retreating army, the cavalry should fight both mounted and dismounted, depending on the terrain and pace of the conflict.

The short-lived Spanish-American War involved little significant action for mounted cavalry, but it inspired Secretary of War Elihu Root to implement army reforms to solve the organizational problems highlighted by the war's conduct.[30] In addition to the Root reforms—which included the founding of the War College, creation of the General Staff modeled after the German Great General Staff, and an improved military school system—the beginning of the new century witnessed a proliferation of American-authored books and articles on tactics and strategy as well as an abundance of English translations of foreign military works by authors who ranged from junior lieutenants to experienced generals. American officers studied recent conflicts, including the Boer War and the Russo-Japanese War, in their reopened and newly founded schools (Leavenworth reopened in September 1902 and the War College in November 1903). Although the education of officers thereby improved, these new professional schools did not adequately consider how dramatically technology, especially the airplane, was already reshaping the battlefield.[31] Instead, motorcycles, bicycles, and lighter-than-air craft drew most of the scant attention paid to new technology.[32]

At the turn of the twentieth century, American officers argued that their cavalry was a multipurpose force able to meet the requirements of various situations. Debates continued to rage, however, over the use of the saber, the relative importance of mounted versus dismounted tactics, the usefulness of various firearms, and the value of shock action.[33] Military and governmental circles constantly debated improvements in size, organization, armament, and doctrine in American versus European models. American cavalrymen recognized the challenges confronting their branch and, without rejecting modernization, remained active in the debates that would determine their future.

Contemporary European military observers acknowledged the unique usage of the American cavalry by refusing to call American mounted soldiers "cavalry" because they were not primarily trained and utilized on the battlefield for mounted combat such as shock action or pursuit of a disorganized or fleeing enemy. Instead, the Europeans called them mounted infantry because they were not "traditional cavalry of the lance- and saber-wielding kind."[34] Captain Moses Harris of the 1st Cavalry argued that Europeans had stigmatized American cavalry "as mounted infantry" because "the methods employed by our cavalry were opposed to old-world traditions."[35] Paradoxically, modern cavalry scholars identified the use of cavalry during the Civil War as "both the apotheosis and the demise of the cavalier tradition that was so dear to Americans of the first half of the 19th century" despite the reality that this cavalier use had never existed in the United States.[36] The European concern about a cavalry that had been "converted to drab mounted infantry, who did their killing with revolver and repeating carbines," speaks to the unique dismounted use of American cavalry.[37] It also demonstrated the contemporary European belief that a real cavalry fights mounted.

The British Cavalry Prior to the Great War

British mounted forces had existed for centuries but lacked even the tentative unity attained by the Americans. Yet British cavalrymen faced many of the same problems regarding professionalization and modernization as their American counterparts. According to Lieutenant-Colonel G. F. R. Henderson, a major British contemporary military theorist, the debate over the cavalry's future was "one of the most interesting and most momentous questions of the day."[38] British cavalry had a far longer mounted tradition than its American counterpart, and this long history affected the British attitude toward reform.

Cavalry organization and roles differed greatly in the United States and Great Britain. From 1880 until after 1918, the British had thirty-one regular cavalry regiments and raised additional irregular forces when needed. Regular units were loosely divided into four classes: household, heavy, medium, and light. The line cavalry, which usually excluded the household class, served either overseas or at home on rotation in the United Kingdom, whereas household cavalry remained in the United Kingdom prior to the Great War to protect the sovereign and the royal palaces. All regular cavalry had swords and, except for the Lancers, who carried a light ash or bamboo lance, carbines for dismounted service. Historically, the light cavalry was used for gathering noncombat information, scouting, and thwarting enemy reconnaissance as well as engaging in skirmishes and pursuing the enemy. Light cavalry also participated in independent actions, including cavalry raids during which large bodies of troops entered enemy territory with the aim of inflicting economic and structural damage without getting involved in a major conflict. The light cavalry's main strength was its speed over long distances, made possible by lightly equipped men on smaller horses. Although the light cavalry was trained to exploit the charge, that role was historically assigned to the heavy cavalry.[39]

The heavy cavalry used larger horses and men (sometimes wearing armor) to break through enemy cavalry and infantry formations in cooperation with friendly infantry and artillery. The uses of medium cavalry were less well defined than those of the other types. The size and weight of the horse and rider fell between those of the light and heavy cavalry. By the late nineteenth century, little separated light, medium, and heavy cavalry in actual use, in part because of the limited numbers of each class.[40]

Complicating matters further was the extensive use of irregular mounted units and volunteer cavalry troops in colonial warfare to compensate for an insufficient number of regular British cavalry troops.[41] These irregular mounted and imperial forces included mostly local volunteers and were intended to be short term.[42] These "yeomanry," armed with swords and carbines, were paid to train for less than one month a year and could be utilized for home service during incidents of civil disorder or foreign invasion, but by the 1880s, they served primarily ceremonial purposes.[43] The number of yeomanry regiments varied from 1880 to 1914, with thirty-eight regiments in 1899 and fifty-seven in 1914.

Throughout the empire, locally raised irregular mounted rifles and light horse units provided a force that was less costly, but less well trained, than regular cavalry units. Mounted infantry supplemented regular cavalry in the First Boer War (1880–81) and in Egypt (1882). When pressed, such units were formed simply by issuing a horse to an infantryman with no new training. Like the regular cavalry units, the titles "mounted infantry" and "mounted rifles" were often used interchangeably, causing confusion. The roles of irregular units were not always easily distinguished from the duties of the regular cavalry, as on occasion mounted riflemen and infantry charged on horseback, sometimes using bayonets as swords or even makeshift lances.[44]

For the regular cavalry to maintain its preeminence over other mounted troops and to prevent their replacement, cavalrymen had to demonstrate that they had something unique and essential to offer that these supposedly cheaper units did not. This uniqueness was mobility and the use of *arme blanche* tactics. The arme blanche held a central position in cavalry tradition and training in Britain. The major difference between regular forces (including the Indian cavalry and yeomanry) and irregular cavalry (including mounted rifles or light horse) was not tactics, but training.[45] Training for the regular cavalry included the stereotypical charge with men closely packed in close order, presenting a solid mass of horsemen to their enemy. Although the maneuver was difficult to learn and carry out, it was quite rightly described by one scholar as "the pinnacle of the cavalry's achievements" and as "critical to cavalry ethos and doctrine."[46] Though the tactic was rarely used, senior cavalrymen, including Henderson and John Denson Pinkstone French, viewed training in it as vital to the morale of the regular cavalry and its cavalry spirit.[47] The charge separated the cavalry from other mounted troops. This was not the only form of charge available to mounted troops.[48] The irregular cavalry was capable of other types of charges, such as those with more open formations, but not the classic form, lacking both the equipment, such as the appropriate heavy mounts, and the training to complete this complicated maneuver.

Ironically, cavalrymen stressed the unique role of the charge when new technologies of modern war called into question the importance of a cavalry predominately trained in this tactic. Military schools and other academic institutions, including the Royal Military College Sandhurst, the Royal United Service Institution, and the Army Staff College Camberley, provided a forum for officers to debate the future uses of mounted forces. While teaching at the Army Staff College Camberley between 1891 and 1899, Henderson and other officers actively debated and tested various possibilities for the organization and use of military branches, including the cavalry, by studying recent wars.

Henderson's writings and teaching at the Army Staff College helped to revitalize the study of the American Civil War and the functions of a cavalry force that fought both on horseback and on foot. His work was a direct response to those who ignored changes in cavalry tactics and who had forgotten the bloody and ineffectual charges of the Franco-Prussian War and other earlier wars. When Wilhelm II, also known as the "uniform-loving cavalry devotee," according to one historian, ascended to power in Germany, military thought in Europe reverted to old theories of employment, ignoring or rejecting the newer dismounted tactics associated with American warfare.[49] Henderson's studies of cavalry action in the American Civil War increasingly lead him to the conclusion that the branch's strength and success was a result of firepower. He utilized his position to encourage the British cavalry to increase its use of firepower and dismounted action.[50] Henderson's arguments, however, were not persuasive enough for many cavalrymen to accept them completely. A passionate debate over the correct primary role for the British cavalry—fighting as mounted infantry or utilizing sword and lance en masse in the arme blanche—waged on throughout the early twentieth century.[51]

While debate proceeded over the future missions of the cavalry, the Second Boer War provided new firsthand experience for the cavalry and became an opportunity to test new tactics. Despite utilizing both mounted and dismounted tactics, the British regular cavalry did not accomplish much. Insufficient troop numbers hampered operations during the war. The British regular cavalry was not large enough even at the beginning of the war to accomplish all of its duties. Mismanagement of these forces caused additional problems. Commanders frequently overextended the limited number of cavalrymen prior to battles. Extensive, usually fruitless reconnaissance missions left them too weakened to provide much assistance on the battlefield. Even when the cavalry was not frittered away on scouting missions, it was ordered out for every alarm during the night, which prevented it from regaining its strength. Both regular and irregular cavalry lacked unit cohesion because they were frequently broken up and assigned to different columns.[52]

The use of British irregular forces to compensate for the insufficient number of regular cavalry units masked British cavalry reforms. These irregular mounted forces included the colonial Imperial Yeomanry and Mounted Infantry, which played a larger role in the conflict than regular forces.[53] The Second Boer War was a mixed experience for British mounted troops, and as a result, officers took various, and sometimes contradictory, lessons from it.[54] Extensive use of irregular mounted forces sparked intense debates about their future. Should mounted infantry and mounted rifleman continue to exist? Could irregular forces perform the duties of the cavalry more cheaply? Did Britain require a regular cavalry force at all? These questions were tied to the larger reform and reorganization of the British Army in the early twentieth century.[55]

Army reforms from 1902 to 1914 addressed fiscal concerns, new technological developments, and the deficiencies of army organization and doctrine that recent colonial warfare had revealed. Historians agree that many of the reforms prior to the Great War were also shaped by fiscal restraint.[56] Among these reforms were the abolition of the post of commander in chief, the creation of the Army Council, a reorganization of the War Office, the creation of the general staff, and the publication of field service regulations. The first common written doctrine for the entire army, *Field Service Regulations, Part II—Administration,* appeared in December 1908, and with the training manual for all arms, *Field Service Regulations, Part I—Operations,* it formed what the Marquess of Anglesey called the "standard rule book for the conduct of a major war."⁵⁷ An additional major reform that helped to professionalize the cavalry was the founding of the Cavalry School at Netheravon House on Salisbury Plain in 1904.⁵⁸ This school, like the schools in the United States, did not end debate over what roles the cavalry could or should fill.

Although it started over the "relative merits of mounted action with the *arme blanche* and dismounted action with the rifle," the "Great Cavalry Debate" (a name bestowed in 1986 by the Marquess of Anglesey, the author of the multivolume work *The British Cavalry*) also questioned whether strategic reconnaissance and protection duties were more important than previously thought.⁵⁹ Anglesey argued that tactically the cavalry remained the "chief means of mobility, while its strategic value in reconnaissance, raids and protective duties" was thought by contemporaries to "increase as armies and battle fronts both expanded."⁶⁰

Those debating the future composition of the cavalry included military theorists, cavalrymen, and soldiers as well as senior British government figures.⁶¹ Every senior British Army officer was expected to have an opinion on this doctrinal issue.⁶² Senior officers such as French, Sir Douglas Haig, Sir Ian Hamilton, Lord Roberts, F. N. Maude, W. H. Birkbeck, Sir Horace Smith-Dorrien, Sir Henry Wilson, and Sir Henry Rawlinson debated the issues in various professional journals, newspapers, magazines, and monographs. Published discussion about the future of the cavalry and technology proliferated between the end of the Second Boer War and the outbreak of the Great War in 1914.⁶³

Advances in weaponry increased the amount of firepower that cavalry could employ, but they also increased the amount of firepower that could be used against cavalry. One possible way for the cavalry to survive against increased firepower was to improve the mounted charge's chance of success.⁶⁴ One possibility was "saddle fire" in which mounted soldiers shot their firearms while remaining in the saddle. Another alternative was for the cavalry to dismount with a firearm and briefly become infantry. Like the Americans, the British tried combinations of the two options.⁶⁵ Arming all cavalry with carbines and training them to fight dismounted would radically alter traditional conceptions of the cavalry. Many members of what Anglesey called the "old school" of cavalry, who supported the continued use of shock action at every conceivable opportunity, such as Lieutenant-Colonel Henry de Beauvoir de Lisle (later commander of the 1st Cavalry Division in the Great War), however, believed that the cavalry could fight well both mounted and dismounted, citing recent colonial campaigns.[66] Although British reformers knew that the American cavalry could fight both mounted and on foot, some doubted that the cavalry could be adequately trained to fight equally well mounted or dismounted.[67] The Earl of Dundonald argued that "theoretical training in both *arme blanche* and riflemanship would in practice mean the perpetuation of the former," because of the historical connection with training the cavalry in the charge.[68]

Much of the discussion differed as a matter of degree not kind, but it nonetheless became heated. A passionate debate grew over whether the sword or the rifle should be the cavalry's primary weapon. Would making the rifle the primary weapon of the cavalry lead cavalrymen to dismount in all conditions, irrespective of the situation, thus making them no longer cavalrymen but mounted infantrymen? Maude warned that "the evil . . . begins when the soldier is taught to rely on the firearm, not on the sword; for then he begins to look on the horse as a mere means of locomotion, and not, as it really is, an essential part of the ultimate cavalry unit.'"[69] Few participants on either side of the debate rejected the future value of either the arme blanche or dismounted fire action, yet they violently contested the emphasis placed on each.

The arguments over how mounted troops should be equipped led to debates over the impact of changes in armament or tactics on the "Cavalry Spirit."[70] Despite the assumption that the battlefield charge would rarely be used in the future, many cavalrymen still believed that the charge was so vital to the cavalry's identity that it must remain the principal tactic no matter how infrequently used. Many traditional cavalrymen believed that the cavalry spirit, defined as the ethos, doctrine, and self-identity of the regular cavalry units, was tied to the charge. Yet Lord Frederick Roberts, commander in chief of the British Army from 1900 to 1904 and often cited as a leader of the "new school" of cavalry reformers by cavalry scholars, rejected that claim in the preface of *Cavalry Training,* the new manual issued in 1904. Roberts wrote, "Instead of the fire-arm being an adjunct of the sword, the sword must henceforth be an adjunct of the rifle." The immediate fierce negative response to this controversial statement forced the Army Council to defer publication of the new manual.[71] *Cavalry Training* was eventually published but without the preface. Roberts refused to back down and published his opinions elsewhere, including in Erskine Childers's *War and the Arme Blanche.*[72]

Officers' opinions on the main cavalry duties varied at the new cavalry school and throughout the branch. Inspector-General of the Cavalry Robert Baden-Powell believed that the cavalry existed to help the infantry win battles, first by destroying the opposing cavalry, second by locating the enemy's main force, third by assisting on the battlefield, and lastly by turning a defeat into a rout. Anglesey broke Baden-Powell's description down further by noting that the cavalry's strategic duties included covering the army's front, finding "the enemy's main force and concealing their own," and "threaten[ing] the enemy's communications and forc[ing] him to waste strength in defending them." Tactically, the responsibilities of the cavalry were "to destroy the enemy's cavalry; to keep the infantry informed and protected; to cut off and hold the enemy; to chip in where required on the battlefield; to smash up the enemy in pursuit or to protect one's own side from pursuit." Henderson, however, stated that reconnaissance was the foremost function followed by threatening "the enemy's line of retreat."[73]

In 1907, Douglas Haig, the Director of Military Training on the General Staff at the War Office, provided "what was, effectively, the official definition of the tasks of the cavalry." It focused not on battlefield roles for cavalry but its support functions: reconnaissance, security, scouting, orderly work, and communications. Haig divided cavalry into the independent, protective, and divisional cavalry. Independent cavalry conducted strategic reconnaissance under direct order of the chief. Protective cavalry provided the "First Line of Security for the Army as a whole." Divisional cavalry scouted for nearby infantry divisions, provided communication services, and acted as orderlies between divisions.[74] A few years later, Haig added classifications of observation and reconnaissance cavalry with the former being stationary and the latter mobile.[75]

Despite these differing definitions of duties and roles, all cavalry officers believed that the cavalry and all other mounted units required more training. During the combined maneuvers of 1903, Roberts reported, "One of the points brought prominently to notice was the want of sufficient training in scouting and reconnoitring—two of the most important duties of the cavalry soldier."[76] This weakness became more apparent during the 1910 annual maneuvers on Salisbury Plain.

This exercise did not simply alert army leaders to a want of cavalry training. It also marked the first military testing of airplanes in Britain for scouting and reconnaissance, when Captain Bertram Dickson of the Royal Horse Artillery flew a Bristol machine.[77] A new technology was challenging the cavalry's reconnaissance and scouting functions, even as mounted forces outside of the regular British cavalry threatened its other roles.

While aircraft began to make progress in reconnaissance and scouting, British cavalrymen remained divided over their vision of their branch's future. The British cavalry's internal problems greatly hindered its ability to fight a united battle against the outside danger of aviation taking over its reconnaissance role, as cavalrymen focused on the seemingly more pressing internal issues of training, tactics, and equipment. Training, tradition, and the popular press concentrated on the more exciting and glamorous *arme blanche*, to the detriment of reconnaissance and despite the attempts of Roberts and some of his contemporaries to shift attention to other missions beyond the charge. Sending small groups of cavalrymen to locate the enemy contrasted poorly with the colorful dress and flashing swords pictured by many inside and outside the military. Reconnoitering, raiding, screening, and the like also did not clearly demonstrate a need for a specialized cavalry. Perhaps only mounted infantry would be needed in the future. Military histories and contemporary articles treated cavalrymen as brave chargers deciding battles, not quiet creepers supplying vital information. The charge was so central to the identity of British cavalrymen and fueled their branch's unity (cavalry spirit) that reconnaissance usually occupied only a secondary role in debates. This fact directly affected how British cavalrymen responded to airplanes. Aviation was thus able to appropriate much of the British cavalry's reconnaissance and scouting functions before many cavalrymen accepted them as some of the mounted branch's most important duties.

Early Aviation Development

Just as their cavalry history and tradition influenced how cavalrymen responded to aviation, their initial experiences with heavier-than-air craft development shaped how cavalrymen reacted. During the early years of military aviation (defined here as 1908–17), aviation and ground forces worked together to study their respective strengths and weaknesses in an effort to produce the most efficient military service. This cooperative relationship was not always apparent to those outside of the military or, in some cases, even to those inside. This lack of visibility fueled debates in the press about the appropriate rate of development for aviation and the amount of appreciation military personnel had for aircraft. The maturation of military aviation proved a gradual and uncertain process that took decades.

AMERICAN MILITARY AVIATION

The birth of American military heavier-than-air aviation followed a few years after Wilbur and Orville Wright's first successful flights in 1903. The U.S. government initially refused the Wright brothers' offer to purchase their aircraft or designs. This brief delay resulted from the unsuccessful and costly experiment of Samuel P. Langley, the secretary of the Smithsonian Institute, whose work on heavier-than-air flight ended with a crash in the Potomac River two months prior to the Wright brothers' first effective flight. This experiment cost the War Department $50,000 and made the government wary of new investments in aviation.[78] According to Lieutenant Frank Purdy Lahm of the 6th Cavalry, who was detailed to aeronautical duty by the War Department in August 1907, President Theodore Roosevelt did not pay much attention to Lahm's suggestion to buy the Wrights' patent. Remarks handwritten by Lahm on his proposals noted that when Brigadier General James Allen called the president's attention to the document, "he threw up his hands indicating it was too long for him to read."[79] The Wrights then tried to sell their product to the governments of Britain, France, and Germany between 1905 and 1908, but they failed to agree to terms with any of the nations despite extensive negotiations.[80]

Although they failed to come to mutually beneficial terms with European governments, the Wrights found the American government more receptive after European nations showed increasing interest in acquiring airplanes. The United States War Department responded by establishing an aeronautical division in the Office of the Chief Signal Officer of the United States Army in July 1908.[81] Following a few demonstration flights, the army purchased and received a Wright airplane in 1909.

The army's first airplane, designated Signal Corps No. 1, flew several times in Virginia and Maryland before it was shipped, along with First Lieutenant Benjamin D. Foulois, to Fort Sam Houston, Texas, for further testing. Foulois had minimal experience with aviation, but that was more than any of his army contemporaries, perhaps excluding Lahm. Foulois had flown balloons and ridden as a passenger in airplanes. According to Foulois, he received simple instructions for testing. Army Chief Signal Officer General James Allen ordered him to "evaluate the airplane . . . take plenty of spare parts—and teach yourself to fly."[82] The logbook detailing the flights and repairs of the twenty-five-horsepower, four-cylinder engine pusher plane with two chain-driven propellers highlights the difficulty of the task assigned.

Early tests consisted of shaky takeoffs and dangerous landings with only brief flights in between. Signal Corps No. 1 crashed frequently, resulting in expensive damage to the plane but fortunately only minor injuries to the pilot. Continuous repairs and alterations to the airplane, however, quickly consumed the $150 allocated by the government for this purpose. Foulois provided an additional $300 from his own personal funds to keep the aircraft in working order. It took a long period of trial and error for this early aviator and his mechanics to learn to pilot and fix their new machine.[83]

American aviation progressed slowly, much to the chagrin of Foulois. Within a year of receiving the first military plane, Foulois had begun testing it as a reconnaissance platform, which he was able to demonstrate during the Connecticut Maneuvers in summer 1912.[84] Despite his successes, the U.S. Army had only six active aviators and fifteen planes by the summer of 1913. All were acquired for the purposes of reconnaissance and artillery spotting. The U.S. Army was unable to increase this force due to meager appropriations. The Secretary of War's request for $2 million in 1912 to build a force of 120 planes to become competitive with European powers yielded a congressional appropriation of only $125,000.[85]

Despite financial issues, airplanes gained an official organization and began to receive recognition in regulations shortly before the 1914–16 Mexican border troubles and the Great War. The March 1914 *Field Service Regulations* included the statement that "in forces of the strength of a division or larger, the aero squadron will operate in advance of the independent cavalry in order to locate the enemy and to keep track of his movements."[86] Four months later, in July, federal law created the aviation section of the Signal Corps.[87] By the time the United States entered the Great War in the spring of 1917, military aviation consisted of sixty-five officers, 1,087 men, and, according to an Air Corps Tactical School publication, "fifty-five obsolete airplanes fit only for training purposes."[88] This material deficiency did not reflect a lack of interest in aviation's potential. American officers continued to observe and comment on the use of airplanes in the Balkan Wars and in the opening years of World War I. The poor equipment and dearth of personnel that plagued aviation before the Great War changed drastically when the United States entered the conflict.

The American social reaction to the airplane was similar to reactions to other new machines of the era. From the late nineteenth century until the 1960s, Americans thought that machines would be a panacea to modern troubles and bring about a better future or even a utopia.[89] Airplanes were one of the most welcomed of these new technologies. The American public's faith in aviation raised it to the status of a secular religion. Starting around 1910, "they worshiped the airplane as a mechanical God and expected it to usher in a dazzling future."[90] In an era when Americans "viewed mechanical flight as portending a wondrous era of peace and harmony, of culture and prosperity," any group opposed to or preaching caution could easily be accused of backwardness and conservatism.[91] The military had a reputation as one of these groups. This fact is essential to understanding many of the debates that followed.

BRITISH MILITARY AVIATION

Unlike the American public, the British public did not believe that airplanes would create a peaceful utopia, but they saw the airplane as a threat to its nation's security by eliminating the historical defense offered by the English Channel. Lord Northcliffe, the great newspaper magnate, summed up the danger when he stated, "England is no longer an island."[92] Having a strong navy meant little if the enemy could simply soar above ships lacking anti-aircraft defenses and reach the mainland. Aviation enthusiasts campaigned to rouse the public and government to the danger of lacking airships and airplanes. The press initiated various scare campaigns including articles that highlighted the danger of German airships flying over Britain, exaggerating the actual danger to the public.[93] While historians have correctly argued that ignorance of the airplane created inflated ideas of aviation's potential in the United States, they did not always recognize that a similar ignorance produced inflated fears in Great Britain.[94]

Despite the predicted danger from the air, when the lead in aviation development shifted from the United States to Europe, it did not go to the British. The British lagged well behind other nations in military heavier-than-air craft. When comparing Britain and France's aeronautical development in September 1911, the eminent British aeronautical journalist Charles C. Turner commented, "Any comparison of the aeronautical work in France and England at the present moment is humiliating for ourselves."[95] The British, however, kept track of the military innovations occurring in Europe, particularly in France and Germany.

The delay in the development of aircraft, an air force, and air defenses in Great Britain resulted from a combination of factors but not from an inability to recognize the value of aviation. Differing opinions held by government officials and the public on the best path to follow hindered development. While civilian aviation enthusiasts learned to build and fly airplanes by copying their contemporaries, some British officials, most notably Secretary of State for War Richard Burdon Haldane, wanted to investigate aeronautical developments more scientifically separate from and independent of foreign developments.[96] The debate, however, was much more than an academic dispute between different philosophies of progress. It also involved questions about the appropriate combination of public and private endeavors and how much the effort should be centralized.[97]

Haldane's desire to create a more scientific method for developing aviation, replete with a technical committee organized to analyze the problems of flight, garnered criticism from contemporaries. Lord Northcliffe, the great newspaper proprietor, desired a more pragmatic development based on the purchase and testing of already functional foreign aircraft.[98] Arthur Lee, chairman of the Parliamentary Aerial Defence Committee, desired to merge Haldane's and Northcliffe's plans to create an integrated approach to aviation research and development that utilized both distinguished scientists as well as "really practical aeronauts." He accused Haldane of being "very much enamoured of," perhaps just short of "hypnotised by," science. Conceding that pure science "is very well in its way," Lee argued that in the case of aviation development, science "is of more value when diluted by a good deal of practical experience."[99] Still others rejected all approaches to aviation development, arguing that aircraft would not be helpful to the military. The most notable member of this small group was General Sir William Nicholson, the chief of the general staff. Nicholson maintained that aircraft of any type would not be of much use to the army.[100] This opinion did not garner support or prevent the continuing development of heavier-than-air craft.

Ultimately, British airplane development included all of the above strategies to a greater or lesser extent. Initially, the military tried Northcliffe's practical experiments with existing aircraft. Tests by aviation pioneers A. V. Roe, Samuel Franklin Cody, and John William Dunne, flying both triplanes and biplanes in 1908 and 1909, resulted in repeated failures. The War Office had hired Cody and Dunne to create an army airplane in secret, but they failed to keep pace with other nations' developments.[101] Haldane, objecting to the very nature of the testing not just its difficulties, declared Cody and

Dunne's investigations into powered flight as not suitably scientific and fired them in 1909.[102] After removing Cody and Dunne, Haldane hired Mervyn O'Gorman, a well-known consulting engineer, to run the Army Aircraft Factory.[103] This hire and new, scientifically based system seemingly only hampered the developmental progress by adding panels and meetings but little practical trial and error. Impatient, the army turned to a private citizen to buy its first two planes in 1910. Although sources are unclear, these planes were most likely Farman "Longhorn" pusher biplanes.[104]

The nation's attachment to airships also hindered British development of airplanes. Financially, Britain had already deeply invested in airship research and technology. This resulted in a continued preference for and concentration on the proven technology of airships.[105] In 1909, the Advisory Committee reported to the House of Commons that while both airplanes and airships were assigned to the army for experimentation, they were to give their "first attention to the dirigible."[106] Some supporters of airships, such as aviation expert and defender of the dirigible Major H. Bannerman-Phillips, praised the "enthusiasm, labour, and ingenuity" spent on aircraft but warned against the temptation to confuse admiration and sympathy for an inventor's "daring and determination" with the value of airplanes. Writing in a service journal, Bannerman-Phillips concluded that airplanes "remain interesting scientific toys, of little or no practical value for purposes of war."[107] On August 3, 1909, eleven months after Bannerman-Phillips's article, the *Manchester Guardian* reported that the House of Commons also believed that "aeroplanes [were] of little use at present" for army purposes. Members of Parliament, however, thought that heavier-than-air craft might be used to conduct reconnaissance once they "achieve[d] higher altitudes and gain[ed] greater reliability and control."[108]

The divided commitment between airships and airplanes manifested itself in February 1911 when the Royal Engineers Air Battalion was established. It consisted of an airship company and an airplane company. The airship unit possessed two airships, and the airplane company initially consisted of five planes of various types.[109] Reflecting the British military's slight preference, pilots were trained first in dirigibles, and then in heavier-than-air craft.[110] Additional organizational changes followed. The Technical Subcommittee of the Committee of Imperial Defense, founded to study the aeronautical situation in Britain, recommended the formation of the Royal Flying Corps, composed of a naval wing and a military wing. In addition to the two wings, the new organization included a central flying school, a reserve, and the Royal Aircraft Factory at Farnborough.[111] Early training, classes, and experimentation with heavier-than-air craft also took place at Farnborough.[112]

After Haldane's scientific route failed to keep pace with other European nations' developments, Britain eventually recovered by following Lee's suggestion of a combined approach to aviation. The extent to which Britain had fallen behind became clear when observing progress in other nations, particularly France. According to British investigations, the French War Office had about thirty airplanes immediately preceding their September 1910 maneuvers. The airplanes utilized during these exercises completed "successful aeroplane reconnaissance" and spurred a doubling of the French airplane arsenal (estimated by the British to be about sixty machines) by the end of 1910.[113] In addition to the number of planes, French aviators had reached extraordinary objectives including cross-country flights of more than 130 miles, nonstop flights of over 250 miles, flights over 100 miles with two passengers, and flight to altitudes between 8,000 to 9,000 feet. Both American and British development at the end of 1910 paled in comparison.[114]

Britain closed the distance between itself and other European nations in aviation development by utilizing the pragmatic tactic of adopting foreign technology and altering it for British consumption. This was especially true when it came to aircraft engines.[115] These developments made it possible for ten Royal Flying Corps planes to participate in the 1912 army maneuvers and thirty-nine planes to appear in the 1913 fall maneuvers.[116] The types of airplanes tested at the 1912 maneuvers included designs by both French (Breguet and Maurice Farman) and British (The Royal Aircraft Factory, Short, and Cody) manufacturers. All airplanes had foreign engines made by Gnome, Renault, or Astro-Daimler.[117] Despite poor weather, these planes conducted air observation that convinced most British commanders of their future importance and provided "many lessons . . . which proved of utmost value when war broke out," according to the official Royal Air Force history.[118]

By 1914, even as Britain was catching up with neighboring European nations in aviation technology, it far surpassed American aviation development, which did not match British advancements until the 1920s. At the beginning of hostilities in Europe, Britain had sixty-three aircraft in the Royal Flying Corps and fifty operational airplanes in the Royal Naval Air

Service with several more at home fit only for training purposes. Although Britain had a higher ratio of airplanes to men under arms than its European counterparts, the reality was that in 1914, it was still outnumbered by ally and adversary alike both in the air and on the ground. France, with approximately 120 airplanes; Russia, with 190 airplanes; and Germany, with 232 airplanes, all outnumbered Britain.[119] In addition to their air units, both France and Germany mobilized three-million-men armies, more than three times greater than Britain.[120]

Despite that size disparity, British aviation made noteworthy strides during the prewar years and advanced more quickly than their American counterparts. Most significantly, the early combination of the science of flight and practical experiments contributed to the successful production of the BE2c. This two-seater reconnaissance plane was an exceptionally stable platform for observation. Unfortunately, its stability came at the expense of maneuverability, making it vulnerable to agile interceptors.[121] Despite the problems experienced by the BE2c in operation, the craft emerged as a challenger to the British cavalry's reconnaissance role.

—

American and British cavalrymen's differing opinions about the appropriate roles and armaments of the cavalry branch and the value and use of airplanes reflected their unique histories, cultures, and experience. The American cavalry was a much younger and more flexible force than the British cavalry. During its short history, it had been utilized mostly as a dragoon force, equally comfortable fighting mounted or dismounted. British cavalrymen were members of a service that had existed for centuries and had built their identity around their use of the knee-to-knee charge. Although they were trying to reform themselves to address modern war conditions, debates remained. Complicating these reforms was the continual need to respond to and, when necessary, to incorporate new technologies. The different experiences with airplanes prior to the Great War also set the stage for how each nation's cavalrymen reacted to the threats and opportunities introduced by airplanes.

The differences between the American and British cavalries were not static; they were continually changing and constantly under discussion. Recognizing the need to adjust to modern conditions, the cavalries of the United States, Great Britain, and much of the rest of the Western world broadly shared ideas across national borders through exchanges in personnel, visits, reprints and translation of articles, and books about the cavalry's future. Examining how commanders utilized the cavalry in previous wars, particularly the American Civil War, the Second Boer War, and the Russo-Japanese War, remained one of the most popular ways of identifying and discussing future roles. Military theorists in Great Britain and the United States gleaned different lessons from these wars, which had a significant impact on how each country planned to use their cavalry in the future and how they responded to airplanes.

CHAPTER TWO

Early Response
to Heavier-Than-Air Flight

Threats in the Press and Cavalry Reaction, 1903–1917

Historian Roman Jarymowycz observed that the "airplane, with the tank, were kindred specters of a revolution that would eventually conspire to remove cavalry from modern war."[1] In 1903, however, the latter did not exist, and the former's technological limitations and lack of defined roles prevented it from immediately replacing the well-established cavalry. The overthrow of the horsed cavalry took time and did not appear obvious or inevitable to many military officers prior to, during, or even after the Great War. Suspicion of the cavalry's demise grew, however, as aviation technology advanced at a dizzying pace between 1903 and 1914. Skeptics in the popular and military presses questioned the cavalry's ability to survive against airplanes, which they believed spelled certain doom for the historic role of the cavalry on the battlefield. Within five years of the first heavier-than-air flight, the airplane had become a threat to the traditional utility and value of the cavalry.

American cavalrymen writing primarily in the *Journal of the United States Cavalry Association* were more critical of aircraft than their British counterparts before the Great War. Yet prior to and through the early stages of World War I, the American and British cavalries' responses to the airplane and its ability to take over cavalry duties demonstrated that they were composed of rationally cautious personnel. Some mounted regulars even welcomed aviation as a replacement for or a supplement to what they considered the less prestigious cavalry functions, once airplanes became developed enough to fill these roles.

Cavalry and Aviation Linked

Early aeronautical research and development in the United States and Britain consisted of more than ensuring reliable operation and breaking distance, speed, and height records. It also included experiments to determine the possible military uses of airplanes in the immediate and near future. The first application for heavier-than-air craft was reconnaissance, just as it had been for lighter-than-air craft.[2] Lighter-than-air craft, which were mostly tethered, had been used in wars since the French Revolution, including the American Civil War, the Crimean War, the Franco-Prussian War, the Spanish American War, and the Boer War.[3] Military personnel had long desired better intelligence. Airplanes could both improve and weaken aerial reconnaissance capabilities. Their speed and maneuverability allowed them to avoid ground fire better than slower-moving (or stationary) lighter-than-air craft. Airplanes, however, were inferior to balloons and dirigibles because they could not hover over a target silently.[4]

Shortly after heavier-than-air aviation became practical, its reconnaissance potential became a focus in both the United States and Britain. Popular, specialty, and military publications commented on this promise. A 1910 American article noted that airplanes would solve the difficulties associated with reconnaissance by "affording the commander a view of events transpiring behind the veil that screens his front."[5] One year later, cavalryman Frank P. Lahm, winner of the international dirigible balloon contest in September 1906 and called "probably the foremost practical aeronaut" in 1909, stated, "Reconnaissance is where air craft will find their real sphere of usefulness."[6] A fellow officer in the field artillery concurred saying that the "most obvious use to which aircraft will be put by the military will be that of reconnaissance."[7] British officers agreed with the Americans. A major of the 17th Lancers predicted that the "reconnoitring powers of the dirigible and aeroplane in the hands of an expert will be fully realised ere long."[8] This potential materialized in the 1912 British Army maneuvers when aerial craft were assigned to conduct reconnaissance.[9]

Because reconnaissance remained one of the valuable tasks accomplished by the cavalry in both the United States and Britain, it is not surprising that aviation and cavalry soon became linked. Both military personnel and civilians understood this connection. American and British newspapers, magazines, and journals published articles speculating about the possibility of aviation replacing the cavalry in reconnaissance and in other related roles such as liaison, scouting, and communication.

In Britain, the conclusion before 1912 was that the cavalry could not be replaced yet. British officers defended the continued necessity of the cavalry for reconnaissance by noting the primitiveness of heavier-than-air machines. Brigadier-General Henry de Beauvoir de Lisle argued in a widely republished lecture that "aeroplanes could not be entirely depended on yet for acquiring information." Attempting to deflate the argument that the cavalry would be replaced, he stated the belief that no one would "wish to see Cavalry reconnaissance abolished" entirely due to aerial reconnaissance.[10] De Lisle conceded that the time may "come when Cavalry would be used more to verify information acquired by air scouts than to procure this information primarily," but that was as far as it would go. Despite any additional aviation developments, he still thought the cavalry would be necessary.[11]

The general consensus was that aviation would support the cavalry in the field as an auxiliary service and not replace mounted forces. Major F. H. Sykes, when commanding the military wing of the Royal Flying Corps, agreed that cavalry and aviation would work together. He reported that the 1912 maneuvers demonstrated that "aircraft will, it would appear, in no way render the services of cavalry useless, but they should save it much unnecessary work." Put simply, Sykes stated the "air service will form a great auxiliary to the other arms."[12] However, he added that once airplanes improved technically, making it possible to perfect aerial reconnaissance, it was possible or probable that cavalry reconnaissance would end. That time was still a long way off. Concerned with the danger airplanes posed to the lives of aeronauts, the *Times* of London maintained that "an army in the field will have for a long time yet to come to depend upon its cavalry for its exact services of reconnaissance and protection" due to aviation's low level of technological development.[13]

In American publications, glowing predictions of aviation's future potential replaced the British periodicals' balanced treatment of the airplane's current limitations and potential cooperative relationship with the cavalry. Because the United States had only limited experience with airplanes, the American press cited those who had direct experience—British and European officers. As early as March 17, 1907, the *Detroit Free Press* reported that Captain John Edward Capper, commander of the Balloon Section at Aldershot, remarked on the seeming invulnerability of these new machines. He believed airplanes "will move fast and be little liable to injury, as bullet holes in the surface will cause but little damage," and that they "will be able to go considerable distance even against strong winds."[14] The *Los Angeles Times* reported that British Major R. S. S. Baden-Powell, author of works on military tactics, contended that future wars would be fought in the air, possibly rendering "armies . . . useless."[15] American observation of European maneuvers also produced comments on the value of airplanes. The French Army maneuvers in Picardy led the *Living Age* to report foreign observers' beliefs that airplanes might rival gunpowder in their revolutionary impact on warfare.[16]

These more promising predictions of airplanes revolutionizing warfare by limiting or entirely replacing ground troops singled out the cavalry. Again, the American press turned to Europe for its sources. As early as 1908, an American periodical had published the opinion of a French aeronautical expert that the cavalry would soon no longer be needed to conduct reconnaissance or scouting missions because an airplane "could entirely supercede cavalry on account of its speed and the possibility of securing more full and accurate observations." The unnamed expert claimed that the airplane would "revolutionize warfare on the land by altering the whole conditions of the problems of 'information.'"[17] The Wright brothers' experience in Germany produced a similar assessment in a 1909 *New York Times* article, which suggested that an airplane could "accomplish more than an entire cavalry regiment as far as locating the position of an enemy."[18] American airplane experiences also garnered attention. Lieutenant Joseph Fichel's trial of airplanes with aviation pioneer Glenn Curtiss on behalf of the War Department in 1910 led him to declare "that a battalion of aeroplane sharp-shooters will supplant cavalry in the army of the future."[19] The editors of the enthusiast magazine *American Aeronaut* concurred with the evaluation that the cavalry would be replaced or face easy elimination. A 1909 editorial predicted that cavalry on horses were so vulnerable to attack from the air that they were "obviously doomed," adding that both horses and army wagons "will fall before one gust of machine fire from above, like as many children's toys."[20]

One of the most sensationalist and pro-aviation articles in a nonspecialist American publication appeared in the *Indianapolis Star* in 1911. It described a disagreement over the future value of aviation, highlighting an increasing tension between aviation and the cavalry. The article contained the reaction of an aviator to a cavalry officer who dismissed the present value of airplanes. It addressed the controversy that followed a speech by Colonel Walter S. Schuyler, described as a noted Indian fighter and military critic, who had said that "the usefulness of the aeroplane in warfare was vastly overestimated" and argued "that the flying man's hope was practically useless in real war." The *Star* reported that Schuyler's statements had aroused a "storm of opposition" among airplane producers and pilots. The subtitles of the article clearly reflected the aviator's position: "Aviators Declare Air Vessels Must Constitute Cavalry of Army," "Horse to Be Abandoned," and "Birdmen Will Scout Territory for One Hundred Miles in Advance." The author provided only a summary of Schuyler's thoughts while including extensive quotations supporting the aviation proponent's arguments. Pilot Charles K. Hamilton responded by noting Schuyler's obvious bias, suggesting his love for the cavalry colored his viewpoint. Hamilton's own possible bias as an aviator was ignored. He attacked not just Schuyler but all cavalrymen as "almost unanimous in decrying the aeroplane . . . as a new-fangled toy that can not be developed into anything practical." Hamilton argued that cavalrymen, like ostriches, "stick their heads down in traditions and decline to see the aeroplane as it really is." Hamilton excused the cavalry's instinctive "prejudice" as stemming from the "first law of nature, self-protection," believing that the cavalry's "present exalted preeminence" will end as "soon as the aeroplane unmistakably demonstrates its complete usefulness in warfare."[21] This article highlighted the most extreme opposing viewpoints possible for cavalry supporters and airplane supporters: the first perspective that airplanes would never amount to a viable military technology and the second that airplanes would replace the cavalry in the near future.

Far more temperate American journalists mimicked their British contemporaries by discussing the abilities of airplanes to assist and cooperate with the cavalry and other military branches rather than replacing them entirely. Yet they still praised aviation's potential. A 1907 *New York Times* piece was typical of this perspective, arguing that while the cavalry was "designed to scout and develop information," infantry and aviation provided "an additional and more complete means of obtaining information."[22] A few years later, in 1912, the same paper reported that the cavalry's usefulness for reconnaissance remained intact because recent maneuvers demonstrated that the "aeroplane is unable to fly low because of its vulnerability" to fire from the ground.[23] Articles from 1907 through the Great War suggested airplanes were an additional tool for observation and not a replacement for the cavalry.[24]

The American popular press was not alone in declaring the airplane a challenger to cavalry reconnaissance. Military journals also addressed aviation's possible impact on the cavalry, both echoing the arguments of the popular press and providing seemingly authoritative opinions for journalists to print. Some army officers claimed airplanes were of equal or greater value in reconnaissance than the cavalry. Only a month before a similar contention was made in the *New York Times,* Infantry Captain John R. M. Taylor argued in the July 1909 issue of the *Infantry Journal* that one airplane was equal to a large cavalry force.[25] He maintained that the cavalry historically was only able to "succeed in ascertaining where the fog lay densest" without providing any of the additional concrete information required by commanders.[26] He believed that the major time investment required to train cavalrymen—at least three years—seemed unwise when the cavalry's "most important function of obtaining and rapidly transmitting information" might soon be placed "in the abler hands of the navigator of the air."[27] According to Taylor, airplanes could do the job better and more cheaply. This cost-efficiency argument echoed earlier arguments by those who favored mounted infantry and mounted rifles over the regular cavalry (see chapter 1). These arguments provoked curt responses from those who took aviation's current limitations into account.[28]

Responses to Early Aviation Predictions Prior to 1912

Coverage of aviation and its connection to cavalry roles in the American popular press did not provoke an immediate cavalry response in the Cavalry Association's journal. Cavalrymen contributed only a few articles mentioning aviation to the *Journal of the United States Cavalry Association (JUSCA),* a major forum for cavalry concerns. Instead, they focused on concerns that appeared more urgent at the time such as the need for a chief of cavalry, equipment and arms for their branch, training, service schools, horse care, and doctrine.

As with aviation, the cavalry branch did not lack advocates. In fact, three Army chiefs of staff in the beginning of the twentieth century came up through the cavalry. However, their tenures expired before military airplanes appeared in maneuvers (Samuel B. M. Young [1903–4], Adna R. Chaffee [1904–6], and J. Franklin Bell [1906–10]), and they were not able to defend the cavalry against aviation's inroads. Regardless of whether these chiefs of staff supported cavalry policy, the ongoing campaign for a designated cavalry chief demonstrated that something was missing. Junior officers felt a lack of leadership and demanded a cavalry chief.[29] Without a strong branch chief to educate the public, Congress, and even cavalrymen about the requirements and importance of the cavalry until after the Great

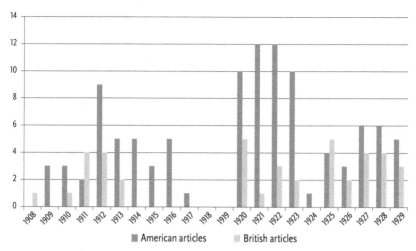

American articles   British articles

FIGURE 1. Articles mentioning aviation in the *Journal of the United States Cavalry Association* and the *Cavalry Journal* (UK). American publication was suspended July 1918–April 1920; British publication was suspended October 1914–April 1920.

War, cavalrymen turned to the United States Cavalry Association to discuss cavalry issues. The fact that the journal rarely covered aviation until 1912 suggests that airplanes were not seen as a serious challenge to the cavalry. Of the more than 350 articles published in the *JUSCA* from 1908 to 1911, only eight mentioned aviation (see figure 1).

The first references to airplanes in the *JUSCA* were a handful of short articles and brief notes in longer articles on various topics, the first of which appeared in 1909. Surprisingly, even as other military officers expressed their opinions on aviation's impact on the cavalry in the popular press and in military journals, such as the *Journal of the United States Cavalry*, the *JUSCA* offered few opinions on the innovation. When the *Journal of the United States Cavalry* published the article "Cavalry and the Aeroplane," the editor of the *JUSCA* was more upset with the *Infantry Journal* discussing the cavalry's business than any predictions the *Infantry Journal* made about aviation's impact on the cavalry.[30]

The scarcity of articles about aviation in the *JUSCA* may have been the result of editorial policy or a lack of interest in the subject among its contributors. American cavalrymen authored few articles concerning aviation until 1912. Nonetheless, the patchy pre-1912 coverage started to indicate to *JUSCA* readers the airplane's potential value and danger to the cavalry. The increase in articles about the cavalry and aviation after 1912 reflects a growing concern among the journal's readership and a desire to learn more about the airplane. To meet the demand, the *JUSCA*, like the American popular press, repeated information from European sources, using the experiences of the British, French, Germans, and others to provide information on aviation capabilities and limitations. These contributions from overseas (including reprints of European articles and American-authored articles based on observations of European activities) focused on the new technology's flaws and limitations, which were not difficult to identify. They included technological constraints due to weather, mechanical limitations and unreliability, and unfavorable comparisons to cavalry capabilities. American cavalrymen were reassured that the unreliability of airplanes meant they could not be depended on as the sole means of reconnaissance.[31]

The *JUSCA* did not condemn airplanes out of hand. It also reported on aviation's positive aspects. In November 1910, it reprinted an article from the British periodical *Broad Arrow* that claimed the airplane had already fulfilled "sanguine hopes" and "achieved more than its most ardent friends anticipated" in its capacity for observation.[32] The article's author contended that the French Picardy maneuvers demonstrated that, unlike the army's present "eyes and ears" (the cavalry), the future "eyes and ears" (airplanes) would "have an uninterrupted view, not only of the enemy's front, but of his flanks, center and rear."[33] The author's faith in the future value of airplanes was not diminished by their current weaknesses, and these weaknesses did not prevent him from declaring airplanes an improvement over the cavalry. Other authorities simply hoped that if some predictions came true, airplanes could lighten cavalry exploration duty, eliminating the need to "pierce the enemy's protective screen and find out the movements of his main columns."[34]

Of the pre-1912 articles mentioning aviation in the *JUSCA*, only one author was an active American cavalryman. Five were reprinted from foreign sources. The other contributors were a member of the Coast Artillery Corps and a retired cavalryman. This divided authorship is shown in figure 2. The lack of articles written by American cavalry officers supports Major Nickolaus Reidl's observation, made in a 1912 *JUSCA* article, that cavalry officers had showed "little inclination in the past to carry on a paper war" over aviation's progress. However, Reidl noted a shift in that year, observing that cavalrymen began combating the erroneous ideas that had begun to

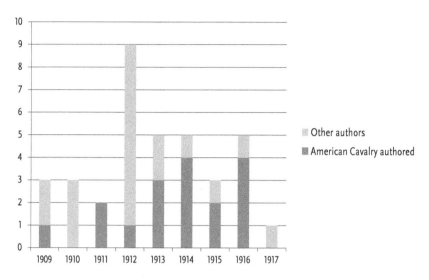

FIGURE 2. Authors of *JUSCA* articles mentioning aviation.

circulate about the overblown value of airplanes and the lack of appreciation for cavalry abilities.[35]

The increasingly positive reports on aviation and discussions in the American popular and specialist press pushed some cavalrymen to address aviation's possible effects on the cavalry, but only indirectly at first. The generally optimistic attitude toward aviation and other innovations led cavalrymen to start warning their colleagues and the public of the danger of overconfidence in unproven new technologies. Identifying himself only as "Boots and Saddles," one writer cautioned that "there is a great danger that a few over-zealous enthusiasts will succeed in getting their hobbies adopted in place of really valuable portions of the present equipment, that have stood the tests of long and hard service successfully." The author attacked the silence of experienced officers because "in only too many instances, these gentlemen are keeping quiet while the faddists occupy the center of the stage and both wings in addition."[36] The United States did not lack aviation faddists in the 1910s with unrealistic expectations bordering on the fantastic.[37] "Boots" was concerned that the reasonable majority was not being heard.

Another writer, a Civil War volunteer cavalryman, echoed this concern that aviation supporters were getting carried away while realists remained silent. He warned that because "the old is [frequently] of more practical value," established technologies should not be abandoned until after a full debate and not simply "because, forsooth, aeroplanes have appeared on our horizon, the old order must give place to something new."[38] Yet he took his argument too far by claiming the invulnerability of the cavalry. The credit-ability of his warning against the dangers of immediate uncritical acceptance of airplanes was damaged when he admitted that he had never seen a "maneuvering aeroplane," but he confidently argued, "A platoon of cavalry could shoot it full of holes with their carbines before it could do any mischief."[39] Such statements in defense of the old were often just as problematic as blind acceptance of the new and weakened the author's argument. It took time for both aviation and cavalry supporters to moderate their expectations before the *JUSCA* could host a meaningful and substantive debate within its pages. The *JUSCA*'s coverage of aviation increased greatly after 1912 as the airplane proved that it could accomplish certain missions in testing and exercises.

Britain before 1912

The British *Cavalry Journal*, like its American counterpart, focused on de-bates over its future organization, mission, and equipment more than on aviation before 1912. The British cavalry had an established leader, but it still lacked a unified position concerning its principal roles. Although the British cavalry had a written doctrine with defined missions, the heated debates concerned its primary role and the primary weapon.[40] This internal debate absorbed a vast amount of attention. Some contributors claimed the pri-macy of the charge, whereas others asserted the importance of dismounted action, reconnaissance, and screening.[41] The last position was not a common one as reconnaissance was not a popular mission. That fact may have been why British cavalrymen appeared more receptive to aviation accompanying or replacing them in the reconnaissance role. Some British cavalry officers rejected the importance of reconnaissance duties for the cavalry. Training, tradition, and popular history concentrated on the more exciting and glam-orous arme blanche. The charge was central to the identity of many cavalry-men and an essential part of how they defined themselves. Reconnaissance lacked the emotional element of the charge.[42]

Some senior officers attempted to end the sometimes vicious debates by advocating for the American policy of a flexible force. In the first issue of the *Cavalry Journal* in 1906, Inspector of the Cavalry R. S. S. Baden-Powell avoided conflict over cavalry use by merely listing the possible roles of the cavalry in assisting the infantry to win battles as "shock action, dismounted action, reconnaissance, protection, long-distance raids, &c."[43] He did not attempt to assign priority to them. He simply urged cavalrymen to keep abreast of new developments to improve their branch's abilities, hailing the American Civil War as a time when "the Cavalry, being newly organised, were untrammelled by old traditions, but were trained and led as common-sense directed, [and] new tactics came into use."[44] He noted the importance of the cavalry spirit and the need to train for all possible roles and missions in a modern war.[45] Yet his article did little to satisfy the debates over the roles of the cavalry or to clearly define the cavalry's usefulness to other military personnel and the larger British public. Five years later, Inspector-General of the Forces J. D. P. French encouraged cavalrymen to be practical because "we live in a time when very divergent views are being expressed in regard to the duties and employment of Cavalry in the field. Some of these views, though put forward with considerable force and ability, appear to be based on theory rather than on war experience and practical training, and if hastily adopted might reduce our mounted forces to impotence."[46]

French published his ideas in the British *Cavalry Journal*, the editors of which wanted it to be the "foremost place in promulgating their [cavalrymen's] ideas," including opinions about "recent scientific developments," such as wireless telegraphy and aeronautics, that were "likely to affect the tactics of cavalry perhaps even more than those of the other arms." The 1911 Preface encouraged contributors to study these changes and to suggest the "most suitable methods of applying those conditions to the accepted principles of cavalry training."[47] Anglesey argued that the founding of the *Cavalry Journal* was "part of what looked like a concerted campaign by the 'shock' advocates" and "was specifically designed to defend and to spread the *arme blanche* gospel."[48] However, the journal also provided articles opposing arme blanche tactics and supporting research into new technological advancements including the airplane.

Both cavalrymen and fliers participated in the early discussions of aviation and its possible impact on the cavalry within the pages of the *Cavalry Journal*. The British *Cavalry Journal* was like its American counterpart in that it published few articles mentioning aviation at first, but it differed by carrying longer articles focused on the innovation instead of just short references. In addition, these articles were produced by British cavalrymen. The proximity of other nations actively testing aviation, especially France and

Germany, made the new technology hard to ignore. The British treatment of airplanes before 1912 was largely positive, arguing that the cavalry would benefit if airplanes assisted in or appropriated its reconnaissance role.

As early as 1909, the argument that aviation could relieve the cavalry from the tedium of reconnaissance appeared in the *Cavalry Journal*. The airplane was seen as a way for the cavalry to increase its mobility by freeing the branch of much of its reconnaissance duties. British Army Motor Reserve officers M. J. Mayhew and G. Skeffington Smyth (formerly of the 9th Lancers) proposed that because mobility was "the *raison d'etre* of Cavalry . . . any device that adds to its mobility increases its power."[49] Airplanes could facilitate the cavalry's work by reporting on enemy scouts or actively harassing them to assist the cavalry in its own reconnaissance, permitting the cavalry to do its jobs better and preventing the cavalry from being wasted, a problem still remembered from the Second Boer War.[50] The airplane's large field of vision, speed, and range could allow aviators to direct large concentrations of cavalry against specific objectives instead of having mounted units "scattered throughout the whole of the strategical zone in the initial stages of operations, as at present."[51] A major of the 17th Lancers proclaimed, "The dirigible and aeroplane have completely changed all conditions, and with a strong stalker's glass in the hands of an expert no body of troops should be able to approach unseen within a distance of three miles."[52] He argued that the new aerial technologies of dirigibles and airplanes would accompany cavalry brigades and divisions at maneuvers and into action in the near future for reconnaissance purposes.[53]

Yet the airplane would not replace the horse, at least not immediately. The continued need for technological development tempered any positive predictions that the cavalry would no longer be required for reconnaissance in the near future. British aviation expert H. Bannerman-Phillips expressed the beliefs of many of his colleagues that airplanes could not "entirely supplant the horseman for reconnaissance, because of their dependence on meteorological conditions," which could "render their employment out of the question just when they are most wanted."[54] Airplanes may prove "excellent auxiliaries," he argued, but under "no circumstances will they ever be able entirely to relieve the cavalry of the duty of reconnaissance" because aircraft could not function at night and could only fly one day out of three on average because of weather considerations.[55] While progress in aeronautics "powerfully affected the public mind," according to a captain of the 3rd Dragoon Guards, these improvements did not justify "the assumption that their use will cause a revolution in the methods of making war in general."[56] A lieutenant of the Royal Engineers Air Battalion concurred with his colleagues, noting the current limitations of airplanes. This aviator argued that airplanes in war would "for the present be very largely limited to tactical reconnaissance," and even in that role they would "in no way replace the cavalry scout, whose capacity for resistance and screening they cannot imitate."[57] He envisioned that cavalry and aviation could work together, with the airplane supplementing the cavalry. The aviator would be responsible for information in "plan" (view from above) while the cavalryman would determine "elevation" (view from the front).[58] The ground scouts would "still be essential in order to supply such information as is unobtainable by the aeroplanes, for the condition of the terrain cannot be ascertained from the air and it will rarely be possible to say whether buildings, woods, &c. are occupied unless they are examined by patrols."[59]

German maneuvers lent support to the views of British cavalrymen and aviators that airplanes could strengthen the cavalry. An observer of the German cavalry maneuvers in 1911 noted that the demands on cavalry for detached duties—to conceal and protect the infantry and artillery—exhausted the cavalrymen, limiting their ability to act when "the decisive moment arrived" on the battlefield. Adding reconnaissance to these other duties would wear out the cavalry and further weaken its fighting ability. Airplanes could vastly reduce the cavalry's strategic (nonbattlefield) duties, including reconnaissance, allowing "its utility as a fighting arm" to be "proportionately increased."[60] A lieutenant of the 19th Hussars and Royal Flying Corps similarly believed that in "some forms of reconnaissance they [airplanes] will probably relieve cavalry to a great extent," thus freeing large forces of cavalry for fighting and protection.[61]

Although some British cavalrymen seemed willing to share, if not cede, their reconnaissance role to airplanes even before 1912, a few seemed concerned that this position could make them vulnerable to arguments that aviation would make them obsolete. Roles other than reconnaissance remained central in many arguments for the cavalry's continued employment, so relinquishing it did not seem problematic. Yet a few cavalrymen appreciated the danger, expressing their concerns in the pages of the *Cavalry Journal*. A lieutenant of the 12th Lancers noted that some cavalry officers were "often too much inclined to impress on our troopers that scouting and reconnaissance are their most important duties," neglecting "their education in fighting tactics."[62] This overconcentration was unwise because airplanes and other innovations (e.g., wireless telegraphy, telephones) would probably replace large cavalry units.[63] If the cavalry became too closely associated with reconnaissance, it might be eliminated if aviation proved capable of taking over the task.[64] A fellow contributor, apparently concerned that officers were putting too much faith in new technologies' ability to usurp previous technologies, argued that "inventions in war tend to neutralise each other," implying that they had weaknesses to exploit.[65] He did not have difficulty finding the airplane's weaknesses. He argued that poor weather made the use of aviation impossible, so "it would obviously be imperative for cavalry instantly to wholly resume those functions of reconnaissance and raid which nowadays may ordinarily be partly performed by aerial means."[66] He would not even concede that aviation could perform reconnaissance satisfactorily in good weather. His position was unambiguous: the cavalry would still be necessary.

Responding to New Technological and Political Realities

From the end of 1911 through the Great War, predictions that the airplane would replace the cavalry proliferated in American publications, supported by data from training operations and real conflicts.[67] As before, the American reports on airplanes' potential were less supportive of the cavalry than British accounts. American newspaper reports, based on French, German, and Italian maneuvers, noted that military airmen believed unequivocally that "for scouting purposes the aeroplane is a most efficient thing and so much surpasses the cavalry, the usual fastest scout, in quickness of obtaining results."[68] A French general argued in mid-1912, "There are no scouts in cavalry, no spies that give better information than the aeroplane about the position and the disposition of the enemy and their available forces."[69]

Such reports finally prompted direct substantial responses from American cavalrymen and the *JUSCA* to aviation's proponents. Many of these reactions centered on deflating the contentions that airplanes could completely replace the cavalry in reconnaissance by both listing the technical limitations of aircraft and the unreliability of the tests performed. *JUSCA* contributors cataloged the shortcomings of airplanes, basing their judgments on European observations of their technical weaknesses, especially the unreliability of early aircraft engines, and their dependence on good weather. One cavalry officer maintained that other officers overestimated the value of airplanes as air scouts based on their observations of maneuvers without considering the lack of actual bullets. In a lecture delivered at the Virginia Military Institute in Lexington, Virginia, and later published in the *JUSCA,* Major Charles D. Rhodes of the 15th Cavalry argued that unless airplanes approached within eight hundred yards of enemy troops, vegetation, smoke, fog, and other conditions would limit the accuracy of their observations. Yet airplanes venturing to within eight to twelve hundred yards of the enemy would likely be put out of action by rifle or shrapnel fire.[70] A 1912 article examining the reliability of airplane motors under fire from the ground, based upon an Italian report of combat with the Turks, supported Rhodes' conclusions. Its author recounted the following harrowing incident that began at the height of six hundred meters:

> A bullet hit the aeroplane; the trail to ascent higher miscarried . . . Captain Montu [a passenger] was wounded and at that very time the motor gave out. Just as the pilot prepared to glide, the motor suddenly resumed its work. Hardly had the former altitude been reached again when the aeroplane was struck by two bullets. The motor did not now work regularly; it emitted a peculiar noise . . . as the motor worked badly, the altitude of 600 meters could not be preserved and, in order to avoid hostil [*sic*] patrol, a material detour had to be made.[71]

The *JUSCA* published many articles penned by foreign cavalry officers describing aviation's limitations. An Austrian cavalry captain argued that aviation faced almost impossible obstacles, including "thick weather, fog, rain and snow, thunderstorms, [and] heavy equinoctial gales" as well as the "great defect" of the "uncertainty of motive power."[72] A major of the 9th Austrian Hussar Regiment agreed that the general opinion expressed in contemporary military literature—that cavalry would no longer "play a decisive role in battle" partly due to the "enormous progress made in fire-arms and aviation"—was mistaken because balloons and airplanes could "assist and supplement" the endeavors of the cavalry only during favorable meteorological conditions.[73] An additional weakness was the airplane's inability to maintain continual contact with the enemy—one of the cavalry's essential functions—for long periods, at night, or during an attack. Authors simply stated that airplanes could not maintain the necessary contact with hostile ground troops, which, unless rediscovered, could threaten friendly positions.[74]

Even as the American cavalry started addressing the possible menace of aviation, a more pressing threat distracted it. The Hay Amendment to the Congress's Army Reorganization Bill of 1912–13 proposed to reduce the number of cavalry regiments from fifteen to ten, the pre-Spanish-American War total. The editor of the *JUSCA* quickly responded, providing the journal's readers with extracts from newspapers opposed to the legislation because "there are hundreds of arguments, legitimate and sound arguments, in favor of an increase of our cavalry and no good reason why there should be a reduction."[75] Fortunately, the measure was defeated in the Senate when Senator Henry Algernon du Pont of Delaware explained that it was removed because it was "inserted without consultation with military authorities," it would greatly weaken the regular army, and the "resulting economies will not compensate for the diminution in the efficiency of our first line of defense."[76]

Although the 1912 attempt to reduce the cavalry was short-lived and unsuccessful, it may have prodded cavalry supporters to pay even closer attention to the possible threat of the airplane. The cavalry could not afford to ignore technological or any other kind of arguments that could be employed to justify its elimination or reduction. The number of articles discussing the cavalry's possible demise in the field of reconnaissance due to aviation increased significantly in 1912. Nine articles appeared in *JUSCA* that year refuting these claims, more than in the previous four years combined (see figure 1). Aviation was no longer a mere oddity to be relegated to sidebars and short articles written by foreign cavalry officers and members of other services. The airplane had become a new tool to be seriously considered by the cavalry's supporters in their debates over roles, strategy, tactics, and planning.

The responses of American cavalrymen to the possibility of losing their reconnaissance function to aviation were similar to the brief arguments by non-cavalry officers that appeared in the *JUSCA* prior to 1913. Most common was the cautiously optimistic belief that airplanes would make a valuable and possibly essential adjunct to cavalry in reconnaissance and scouting. The strengths of one made up for the weaknesses of the other, with both working, as Colonel Hamilton Hawkins of the 3rd Cavalry stated, "hand in hand." Each assisted the other in the completion of their combined task, with neither supplanting the other.[77] A 12th Cavalry captain admitted that airplanes would be of great value along with other units.[78]

In these arguments, neither airplanes nor cavalry took the lead, and therefore, the cavalry remained necessary. The use of airplanes in reconnaissance, cavalrymen argued, did not make the cavalry superfluous. A major of the 5th Cavalry maintained, "There is an ample field for both and that one will often succeed where the other fails."[79] Another officer went further, contending that in the future cavalry would only be able to complete its mission in combination with aerial reconnaissance.[80] A 9th Cavalry major stated flatly that "the aeroplane may become an adjunct, indeed has already become one, but no military man can see in it more than an adjunct—and least of all a substitute for anything that armies have always found necessary." He maintained that cavalry would, "in common with its sister arms, still be called upon to perform in the identical way, the same functions that it performed under Stuart, Buford, Sheridan and Wheeler."[81] Major Rhodes proclaimed that the airplane would be "a very powerful aid to reconnaissance" with the important caveat "when developed."[82] He called the idea that airplanes would replace the cavalry "such an absurdity as to hardly merit consideration."[83]

In their early understanding and acceptance of the new aviation technology, American cavalrymen were not unreasonable. Their opinions were supported by members of the Signal Corps, the branch in charge of aviation. A Signal Corps captain echoed the cavalry's belief that aerial reconnaissance should supplement the cavalry's scouting role, arguing "an ideal army would consist exclusively of cavalry and horse artillery, and the necessary auxiliary troops." The cavalry was "eminently fitted" for screening and reconnaissance, roles that could not be performed by any other service. The Signal Corps would provide only "an auxiliary means of aerial reconnaissance," not a replacement.[84]

Although most of the *JUSCA* articles accepted the assistance airplanes could render cavalry reconnaissance, they also emphasized that reconnaissance would never be completely ceded to aviation because its technological limitations were unlikely to be remedied. Despite flying machines' continual technological improvement, they were still unable to solve the problems enumerated in pre-1912 articles. American objections to the airplane were essentially the same as those expressed by the British. Airplanes remained auxiliary because they could not function in poor weather and certain terrain.[85] One cavalryman took the problem of meteorological conditions to the extreme, maintaining that stationary fogs would prevent flying in the fall, winter, and spring.[86] Even if aviation technology improved to allow flying to occur year-round despite stationary fogs, aerial reconnaissance would not be possible.

Not surprisingly, American cavalrymen attacked the exaggerated claims of aviation's proponents who stated or implied that the cavalry was no longer necessary. A 1st lieutenant agreed with his fellow cavalry officer that aviation enthusiasts had let their imaginations run wild and were leading the public astray. He criticized the military personnel who, "with childlike faith . . . in flights of fancy too vague for sane minds to record, have relegated cavalry to the land of Skidoo." This officer, obviously well-educated, was not above making his point through sarcasm, writing that

> the herald and the porte-crayon of this body-lunatic will picture for you a winged centaur traversing the heavens in a chariot of fire, disdaining the elements, while from heights too great to be reached by any projectile, death dealing missiles will be showered down upon the unsuspecting heads of our adversaries. They will show this Pegasus of the skies performing all the essential duties of cavalry in such style as to make the redoubtable Stuart, Phil Sheridan and Forrest, with their hordes of heroes, look to the future student of history like the proverbial "thirty cents:"—(10 cents each).

Despite his abusive rhetoric, he maintained that he was "a friend of the aeroplane," motivated by concerns that the "waves of popular enthusiasm" would "crash upon the rocks of cold fact" due to the dangers of exaggerating aviation's possibilities.[87] His belief was that the airplane was still only an adjunct to the cavalry, unable to fulfill the capabilities attributed to it. Another cavalryman noted that articles in the daily press had been either too brief in their coverage of aviation or "so manifestly the result of a reporter's imagination as to be untrustworthy."[88] In both these cases, the authors suggested a more practical approach for evaluating aviation's current abilities.

Maneuvers

Practical evaluations of military operations and maneuvers confirmed American cavalrymen's beliefs that airplanes could not entirely replace the cavalry in reconnaissance roles. For example, in the 1912 Connecticut maneuvers, the airplanes' scouting and wireless communication equipment failed to live up to aviation proponents' expectations. A field artillery captain reported that some military officers had the "erroneous impressions" that "the aeroplane could supplant the cavalry in scouting work and reconnaissance," but he noted that the trials revealed many reasons why such thoughts were premature. The captain listed numerous problems hampering the effectiveness of aviation including the dark, heavy fog, hazy atmosphere, dusty weather, poor landing grounds, and weak or unreliable engines.[89] He maintained that the maneuvers demonstrated that "the aeroplane was merely an adjunct to the cavalry" because it was not yet developed enough "to be yet accorded any fixed and definite role upon the battlefield."[90] Reports from trials in Rheims, France, noted that airplanes did "not yet satisfy military requirements."[91]

The popular press began reporting more temperate evaluations of the value of the cavalry and aviation after actual military tests of airplanes revealed the latter's weaknesses. They admitted that military operations and maneuvers had not shown the cavalry to be superfluous, as they had predicted previously, and that cavalry remained invaluable when conducting reconnaissance. For example, a December 5, 1912, article in the *New York Times* maintained that the cavalry was still useful after observing the lessons of the Italo-Turkish War. The writer noted that the airplane reconnaissance results were "not as sensational as has been reported in regard to some European manoeuvres." He also refuted the much-published idea that the cavalry was no longer needed in light of aviation developments.[92]

Although the caution expressed by cavalrymen was shared by members of many of their sister services and the experiments with aviation seemed to support their warnings that most aviation predictions were exaggerated, some aviators attacked the army for being shortsighted and slow to support the continued development of airplanes. One of them was Benjamin Foulois, the most experienced American military aviator. He noted the development of aviation in the United States and chided the public for its lack of support in light of many significant technological developments. Foulois claimed that both the army and public showed a "lack of enthusiasm [for] and active interest" despite maneuvers in Europe and the United States that had demonstrated that aircraft were "here to stay." Admitting that his statements may "seem to be exaggerated and visionary," he had "every confidence in the wonderful military future of these new weapons of warfare" because of his four years' experience with airplanes. He repeated his predictions that "information relative to the operations of the enemy," previously collected by the cavalry, will "hereafter be obtained by the aerial fleet and transmitted to the commander-in-chief long before the invading cavalry has gained touch with the enemy." In contrast with recent unenthusiastic appraisals of European maneuvers, Foulois contended that recent maneuvers showed that "a modern military aeroplane, equipped with a radio-telegraph set, could easily reconnoiter the same area [as the cavalry] in one-half the time."[93] Foulois regarded anyone who preached caution or argued for limited aviation budgets as backward. Some reports supported Foulois's opinion. On April 12, 1914, a *New York Times* article stated Signal Corps tests with airplanes in San Diego were "accurate in every detail" and "report[ed] in fifty-three minutes what it would have taken cavalry two days to find out."[94] In addition to Foulois, General James Allen also encouraged the "country to get busy" and acquire more airplanes. Supported by congressmen, Allen, not surprisingly, promoted bills designed to increase monetary support of aviation development.[95]

Cavalrymen continued to deny that airplanes would ever make their branch obsolete, but some argued that airplanes could relieve the cavalry of several duties, allowing it to focus on its other more important and necessary roles. American cavalrymen echoed the contention made by foreign authors and non-cavalrymen before 1912 that airplanes strengthened the cavalry. As a coast artillery officer stated, "Cavalry in modern war has all it can do, and its burden or work is constantly increasing, consequently, every means of relieving it as much as possible must be resorted to, and the aeroplane promises to be one of the most effective means."[96] Airplanes would spare the cavalry from "onerous reconnoitering duties," thus increasing the cavalry's ability to be utilized on the battlefield in greater numbers.[97]

As aircraft became more reliable and capable, more American cavalrymen began to consider the possibility of conceding the reconnaissance role to aviation altogether. However, like those before them in the *JUSCA*, these cavalrymen attempted to turn this loss into a victory by arguing that ridding themselves of this duty would strengthen the cavalry by freeing it up for other important missions. They denied that relinquishing reconnaissance made the cavalry obsolete or diminished its importance.[98] A cavalry colonel emphasized that "modern inventions relieved the cavalrymen from many hard rides" but not "the necessity of his services."[99] A brother officer praised the ability of airplanes to save horses for combat.[100] A cavalry major also contended that cavalry equipped with its own airplanes would add to the "reconnoitering and combat value of the mounted arm."[101] Cavalrymen were not alone in arguing that aviation strengthened the cavalry. A senior army officer, General James Parker, argued that the airplane had "added to the value of cavalry" for fighting against infantry.[102]

War

In the months leading up to the outbreak of the Great War, cavalrymen in Great Britain and the United States remained uncertain as to their role in reconnaissance and modern war in general. Would the cavalry facilitate airplane reconnaissance work, or would airplanes serve the cavalry? Would the cavalry have a continued role in reconnaissance? The outbreak of conflict, however, forced the British and American cavalries to set aside debates and get to the business of fighting. Mobilizing and training forces, planning campaigns, and engaging in combat took precedence over theoretical discussions about how the successes of aviation in maneuvers translated into actual war and the cavalry's continued value. The British even ceased the publication of military periodicals, including the *Cavalry Journal*, after Britain entered the Great War in 1914, removing one of the major platforms available to cavalrymen to discuss cavalry topics. However, discussions continued in the United States until it entered the war three years later.

American newspapers and journals made up for the lack of aviation and cavalry coverage in Britain by publishing articles penned by observers in Europe. War correspondents reported events and discussions from 1914 to 1917, measuring the value of aviation and the cavalry. Usually, cavalry did not do well in comparison. In the first months of the Great War, a French corps commander declared that a single airplane was "worth as much as a division of cavalry" in reconnaissance.[103] Airplanes could prevent surprises, as "few important movements of troops have been made which have not been promptly reported by aerial scouts before their completion."[104] Indeed, the *St. Louis Post-Dispatch* reported that airplane scouts were relieving the "always overworked cavalry . . . discover[ing] in half an hour what a detachment of cavalry might fail to find in a day."[105]

During the years of the Great War when the United States remained a neutral nation (if only in name), U.S. newspapers continually reported that "cavalry has given way to it [the airplane]"[106] and "armies and navies are useless in modern warfare" unless assisted by airplanes.[107] An April 1917 *New York Times* article had the subtitle "Airplanes Replace Cavalry as Eyes of

Fighting Forces"[108] while the *Boston Daily Globe* quoted Rear Admiral Robert E. Peary's more damming prediction that airplanes would "do the work of cavalry, infantry and artillery combined."[109] The *Kansas City Star* proclaimed the "Cavalry's Last Days," noting that the American army preparing for service in Europe contained no cavalry. It argued that "from the French word for horseman came [the] word chivalry," but with the coming of the airplane, "chivalry ha[d] taken to the sky" and that, according to French army leaders, "everything that cavalry ever did" could "be done better by airplanes."[110]

Nonetheless, American cavalrymen attempted to demonstrate that the cavalry remained a vital element of the military. To do this they utilized reports from the war to attack those who wrote the cavalry was no longer needed. Cavalrymen even requested funds and additional personnel, claiming that the cavalry only failed in the war because it did not have the necessary material to fight effectively. One cavalry lieutenant took offense to reports and "sensationally illustrated articles" that focused wholly on air combat, accusing these publications of hiding the importance of less glamorous missions, such as cavalry reconnaissance. He maintained that the reason the cavalry was believed to have accomplished little in the war was because no one studied the cavalry's actions. Attempting to correct this oversight, he argued that the cavalry continually conducted close reconnaissance when hostile aircraft or weather grounded friendly airmen.[111] He did not endeavor to minimize the value of aviation but simply wanted the cavalry to get the credit it deserved.

This lack of coverage was understandable and perhaps reasonable. Lieutenant Colonel John Stuart Barrows wrote that because cavalrymen spent most of their time in the trenches like other soldiers, traditional cavalry roles such as charges and reconnaissance were "not given much space in the current news." He was clear, however, that the cavalry was not useless or obsolete.[112] Former cavalryman Henry J. Reilly, a military contributor to the *Chicago Daily Tribune* and the *JUSCA*, also wrote to correct what he saw as the "most prominent" erroneous idea: "that the day of cavalry is past." Although Reilly admitted that mounted actions would be "few and far between," he argued that all cavalries would be armed with rifles allowing them to fight like infantry when not mounted.[113]

A 3rd Cavalry veterinarian also pointed out that additional predictions and beliefs about the war did not hold up upon deeper examination.[114] He responded to the predictions that aviation would be cheaper than the horse cavalry and that the war would pit machines against one another. He argued that little of the "phantastic idea" of the predicted "contest of machines" had materialized. He curtly rejected the predictions of a horseless and almost manless war by observing that the conflict had not played out as people had anticipated. Although he focused primarily on automobiles, he also noted the important gap between the predictions about airplanes and their actual abilities. His assessment of the automobile as an excellent auxiliary but "as a substitute for the horse it proves a delusion" seems equally applicable to the cavalry's views of airplanes. He claimed it was the novelty "of the machines and not their economy that made them popular with armies at the beginning of the war." He provided evidence for his contention for the lack of savings produced by machines by citing the average life span of the various war-making components. Using statistics based on the first months of the war and published in December 1914 in the French newspaper *Figaro*, he found that "the average life of a man in this war is *six and five-sixth* days, and that of a horse *four and one-third* days; aeroplanes and automobiles lasted *three* days, and motor truck less than *one* day."[115]

Even as European cavalry in the war seemed to show cavalry to be superfluous, American cavalrymen urged the public not to equate their value to that of the poorly performing foreign cavalry. This partially stemmed from the belief that the latter had entered the war with bad ideas about how to use the vital cavalry arm.[116] Unlike its contemporaries, the United States did not expect to use the cavalry in great charges; instead, as Hawkins argued, the "great work of modern cavalry would not be done in the limelight, that it would be non-spectacular and silent, and that of all the branches of the army it would make the least noise and be the least observed."[117] The fact that European cavalry did not successfully conduct charges in the Great War, therefore, did not affect the utility of American cavalry. Hawkins contended that the press had failed to cover the cavalry's dismounted and scouting work properly because its useful reconnaissance work was not easily observed and thus "unheralded."[118]

Colonel Hawkins regarded popular writers who minimized the cavalry's future as uninformed and "very dangerous when they write of serious subjects about which they know little." He cited Harvard Professor of Mathematics and Chemistry Charles William Eliot as one of those who claimed "cavalry is a thing of the past, that aeroplanes have completely displaced cavalry for reconnaissance." In reality, argued Hawkins, "aerial reconnaissance accomplished much, but, until the trench warfare began, it did not do as much as some people believe." When the armies still moved, the cavalry performed a "great and invaluable service in both screening and reconnaissance."[119] He argued that "aerial reconnaissance alone was not reliable or sufficient" and that aerial and cavalry reconnaissance "should work together, hand in hand." He concluded, as had his colleagues, that airplanes provided a valuable adjunct to the cavalry, making its duties easier.[120]

Members of other branches also defended the cavalry's continued usefulness. A captain of the field artillery (though previously a 1st lieutenant in the 15th Cavalry) authored a book on American preparedness for war motivated partly to correct the "most erroneous belief" that the cavalry's day had past.[121] The author cited an Austro-Hungarian general who argued that because the cavalry's chief asset remains mobility, "however much the aeroplane might replace cavalry in reconnaissance work, this would not affect the value of cavalry, because . . . mobility makes cavalry especially valuable."[122] The cavalry could still fight, and reconnaissance was not its only use. In April 1917, the same cavalry officer argued, "the gathering of information before, during, and after combat, is primarily the function of the aviation service," but in many cases ground reconnaissance remained necessary.[123]

Military reports from the Great War did not support the claim that "aeroplanes had practically supplanted cavalry in reconnaissance" and that the conflict was not "much of a war for cavalry."[124] By mid-1915, military observers abroad rejected the prophets of doom "who pronounced the death sentence of the mounted arm and denied it any but a small, special place as an assistant in battle."[125] Instead, reports noted the indispensability of the cavalry for reconnaissance along with the frequently mentioned belief that the cavalry's glory days were in its future.[126]

Prior to the United States' entry into the Great War, American cavalrymen did not have to depend entirely on foreign conflicts for examples of how aviation and cavalry complemented or conflicted with one another in real combat situations. In March 1916, President Woodrow Wilson ordered the Punitive Expedition under the command of John J. Pershing to Mexico in response to a raid on Columbus, New Mexico, by Mexican revolutionary Pancho Villa, in which fifteen American civilians and soldiers were killed.[127] This, the United States' first combat test of the airplane, supported the cavalry's belief that airplanes remained insufficiently developed to seriously challenge the primacy of cavalry reconnaissance. In describing the brief involvement of aviation, one scholar concluded that the "army's earliest experiences with airplanes were long on promise and short on performance."[128] The flimsy and unreliable Signal Corps aircraft performed some reconnaissance and carried messages, but crashes and maintenance problems quickly deprived the army of aerial vehicles.[129] A year prior to the United States' entry into the Great War, the cavalry was still needed for reconnaissance because the "army's aviation equipment was both deficient and dangerous."[130] This expedition, less than a year before the United States entered the Great War, was not a great demonstration of the airplane's potential value.

In both the quality and quantity of aircraft and engines, the Europeans were far ahead of the Americans during the Great War. However, when the British initially entered the conflict, British aviation was not ready (although it was more prepared than the Americans would be in 1917). When the conflict began in August 1914, the Royal Air Corps sent four squadrons composed of fifty planes, all with French engines and almost half constructed of foreign designs, to France.[131] However, British aeronautical development accelerated once Great Britain entered the war.[132]

When the United States finally entered the war in April 1917, the debates over the relationship of airplanes and the cavalry declined drastically as the focus became fighting the war. The *JUSCA* ceased publication just as its British equivalent had done three years earlier. The United States entered the war unprepared, "especially dependent on foreign sources for artillery, ammunition, tanks, airplanes, and machine guns."[133] The larger American experience of the war was "basically one of learning by doing" because "there was no systematic effort beforehand to assess the new weapon, determine needs, develop a doctrine, and train troops and commanders in its use."[134]

In preparation for the increasing possibility of entering the European conflict, the National Defense Act of 1916 added ten cavalry regiments to the fifteen already existing. Two were formed and remained horse cavalry regiments during the war, but the rest, the 18th–25th Cavalry Regiments, became field artillery units. Only part of the 2nd, 3rd, 6th, and 15th Cavalry Regiments went overseas. They did not participate in much fighting. Most cavalry regiments remained at home along the border without gaining wartime experience. Few cavalry soldiers actually reached Europe, and no cavalry horses were shipped with American troops.[135] Additional debate over the value of the cavalry compared with aviation would have to wait until after the end of hostilities.

—

American and British cavalrymen from 1903 until the outbreak of the Great War responded to the potential loss of roles and missions to aviation differently due to their unique assessments of the importance of reconnaissance, the treatment of aviation and the cavalry in the popular press, their dissimilar experiences with aviation, and their individual experiences of involvement in military conflicts. The American cavalry responded less quickly to a more negative American press, whereas the British cavalry immediately engaged in a discussion about how aviation could help the cavalry. Yet the British lacked a consistent and unified position and remained mostly unconcerned about the possibility of aviation usurping its reconnaissance duties because they considered them to be a low priority. Individuals within the American and British cavalries addressed the real-world challenges of technologies in transition when no one was certain what airplanes could actually accomplish in the present or near future.

Although skeptical of aviation's revolutionary impact, the *JUSCA* and the *Cavalry Journal* revealed British and American cavalrymen to be primarily cautious technological examiners. Rather than rejecting new innovations that might expropriate their roles, American and British cavalrymen were cautious. They can be seen as rational people responding to the uncertainties surrounding the limited but rapidly evolving capabilities of military airplanes prior to the Great War. Although they had little experience with the new technology, the cavalrymen saw it as having far too many limitations and drawbacks to fulfill the aviation supporters' predictions that the cavalry would no longer be needed to conduct reconnaissance. Many, however, welcomed its ability to add to the cavalry's capabilities by providing additional information and relieving it of time-consuming tasks. Certainly, the new technology had not entirely appropriated the reconnaissance missions of ground units. Unfortunately for the cavalry, there was a growing impression among the public and politicians that modern war and airplanes had already made them obsolete.

The horse cavalry did not become obsolete overnight, nor did aviation instantaneously fulfill the predictions made for it. The horse cavalry continued to play a vital role in the militaries of the United States and Great Britain well after the invention of motorized vehicles. The caution shown by cavalrymen and government officials in both countries about aviation prior to the Great War was a skeptical but rational response to an unproven innovation. It would take time for aviation to become the revolutionary technology many claimed it would be.

CHAPTER THREE

Developing a Relationship in the 1920s

As occurs with almost every conflict, military theorists and officers, politicians, and the media began debating the lessons of the Great War before it ended. The success and potential value of the technologies, strategies, and tactics employed came under heated debate. How innovations—such as tanks, radio, gas, and airplanes—should be used in the future became a contentious issue. Whether the cavalry still had relevance was another subject of debate. British and American cavalrymen defended their branch against claims that technology had made the horse-mounted soldier obsolete, citing examples from the war to argue that the horse cavalry had not outlived its usefulness despite improvements in the capabilities of airplanes and other technologies.

Once again cavalrymen argued that aviation, in cooperation with cavalry, was necessary for military success. However, British and American cavalry forces had different problems arranging that cooperation because the United States and Great Britain organized their air arms differently. The British cavalry had to contend with an independent aviation organization, the Royal Air Force (RAF), whereas American cavalrymen had an air force within their own service, the United States Army. In both nations, however, cavalrymen demonstrated a desire to apply the lessons of the last war to strengthen their organizations by working with aviation units more intimately. This manifested itself in the form of demands for combined training to learn how cavalry and airplanes might support each other as well as requests for organic aviation within cavalry.

The success of aviation during the Great War, combined with the seeming absence of the cavalry from the fighting, led many outside observers to resume the prewar argument that the airplane would supplant the horse. Cavalrymen, long used to such charges of irrelevance, generally responded with dignity, moderation, and reason to those calling for disbanding the cavalry. However, the strain of defending their branch's utility must have taken its toll, as some cavalrymen began resorting to emotion rather than reason in their responses.

British and American cavalrymen continued to utilize journals, especially their nation's respective *Cavalry Journal*, to discuss their opinions and concerns. This discussion expanded as familiarization with airplane technology grew. Figure 1 (chapter 2) demonstrates that the number of cavalry journal articles discussing aviation in 1908–17 and 1920–29 increased from thirty-six to sixty-nine articles in the American periodical and from twelve to twenty-nine in the British journal.[1] Some of this increase may be attributed to a rise in the average number of articles per issue in the 1920s, but a slight increase is still distinguishable. In the case of the U.S. *Cavalry Journal*, the increase was from slightly over 4 percent to more than 6 percent of articles (a 50 percent rise). The real significance, however, was not the total number of articles mentioning aviation, but the quality of the articles' treatment of the topic. Historian William Odom has stated that only three U.S. *Cavalry Journal* articles from 1915–25 "explore the relationship between the air service and cavalry."[2] His total is only accurate if articles mentioning both aviation and cavalry in the title are counted. This calculation misses dozens of articles that address the relationship without announcing so in their titles.

This chapter will provide a summary of the discussions of the air forces and the cavalry in both the United States and Britain following the war as found in the cavalry journals as well as in newspapers, reports on maneuvers, and doctrine. It will begin with a brief description of the status of the airplane–cavalry relationship at the end of the war and go on to describe the relationship between the cavalry, aviation, and modern war during the 1920s. The old argument that the cavalry was obsolete would not die; in fact, it was apparently strengthened by the experience of the war. Cavalrymen continued to defend their existence, updating earlier arguments to take new technological developments into consideration. They dusted off the prewar arguments examined in chapter 2, including their warnings about overconfidence in technology and the operational limitations of airplanes. Yet despite aviation's shortcomings, cavalrymen also understood that it would be of considerable service to them once it had matured. To expedite this process, the cavalry and the air forces worked together in maneuvers and border operations to identify the fundamentals of successful cooperation, though this occurred far more often in the United States than in Great Britain.

Postwar Status of Debates

Rather than examine the cavalry's performance in the war, which would fill the pages of multiple volumes and actually has done so, this chapter focuses on the relationship between the cavalry and aviation during the Great War and the following decade.[3] Not surprisingly, reconnaissance served as the primary link between airplanes and the cavalry, just as it had before the war. The American Expeditionary Force (AEF) Cavalry Board, created in 1919 to determine what lessons the cavalry might draw from the war, reported that the greatest change following the war would be the transfer of responsibility for strategic (long-range) reconnaissance from the cavalry to aviation, leaving the former free to focus on tactical (short-range) reconnaissance operations.[4] The AEF Superior Board on Organization and Tactics—created to review the war for lessons for all American forces—agreed with the Cavalry Board that strategic reconnaissance would be primarily assigned to airplanes, but with the understanding that the cavalry remained necessary for near-strategic (mid-range) and tactical reconnaissance when actual contact with the enemy proved necessary.[5] Its report combined the findings of several postwar boards established to assess the war's lessons in late 1918 through the middle of 1919, including the aforementioned Cavalry Board.[6] According to the cavalry major who served as the assistant chief of staff for military intelligence, aviation's ability to "see over the hill" and determine enemy dispositions behind the front lines as well as take photographs for commanders had transformed the collection of strategic information but did not replace the cavalry's tactical role.[7]

Despite these findings, the postwar popular press continued to denigrate the usefulness of the cavalry after the war, relying on many of the same arguments it had used before the United States entered the conflict. The popular press adopted an attitude reminiscent of one expressed in a prewar congressional debate over appropriations for war preparedness. The ensuing debate pitted those who believed that the cavalry had been replaced by aircraft against those arguing that cavalry had never been so valuable.[8] Proponents of the former argument included the press and were not difficult to find in the United States or Britain prior to the war (as explored in chapter 2). Popular newspapers and magazines, such as the *St. Louis Post-Dispatch* and the *Washington Post,* declared the cavalry obsolete because "aeroplanes [now] serve as the eyes of the opposing armies," so the airplane service can "supplant the cavalry" for reconnaissance.[9] British General Lord Horne, the only British artillery officer to command an army in the war, reported hearing this same conclusion on many occasions during the last two and a half years of the Great War: "'The day of cavalry is past; cavalry is doomed: cavalry fulfills no good purpose.'"[10]

After the war, the press repeatedly reported the belief that the cavalry was no longer useful in modern war, primarily due to technological advancements. New inventions, such as machine guns, tanks, gas, and airplanes, helped to create a landscape characterized by trenches, which curtailed the cavalry's mobility, its primary characteristic. A letter to *The Times* of London proclaimed that cavalry operations were impracticable as they would be observed and bombed from the air. Furthermore, tanks could do everything the cavalry could do at least "as effectively" and far more cheaply than horses.[11] A French cavalry captain, who was also the air service attaché in the French Embassy, argued poetically that the mounted cavalry fights of old had no place in modern war, "where death flashes from thousands of points; where battles are won or lost without adversaries even coming into contact with each other; where, despite the greatest precautions, the losses are so immense."[12] *The Nation & the Athenaeum*, a weekly British liberal political newspaper, reported that the "experience of the war showed that cavalry are as obsolete as bowmen."[13] The media's primary conclusion in the decade following the war, as *Current History* reported in 1928, was that the "cavalry [was as] dead as a dodo" because of the siege-like nature of trenches.[14]

Cavalrymen as well as other military officers rejected these findings, yet the limited use of the mounted branch during the war complicated the horsemen's efforts to provide proof of their continued viability. This was especially true for the American cavalry because the United States' late entry into the conflict, combined with the low priority accorded the transport of horses to Europe, meant only a small cavalry contingent reached the front and did so without its mounts. The portions of the 2nd, 3rd, 6th, and 15th Cavalry Regiments that made it overseas while still retaining their identity as cavalry units served primarily as military police and guards, although they were also assigned remount duties.[15]

After the war the AEF Cavalry Board commissioned a survey of American cavalry members who served during the war, which provides a good picture of the frustrations experienced by their branch. A sergeant of the 4th Cavalry reported joining the unit to get the chance to travel. He got his wish, but his usual travel assignment consisted of driving mule teams from Arizona to Texas, which was not exactly what he expected; he never went overseas.[16] An 11th Cavalry Regiment sergeant who also saw no service outside of the United States despite being scheduled for overseas duty several times moved horses by train to Fort Meyer, Virginia.[17] Members of the 12th and 306th Cavalry Regiments related similar experiences.[18] Even those cavalrymen in the 15th Cavalry Division who went overseas had little to say about their experiences. A private summarized their experiences best, reporting that they left their horses behind with the promise they would be on the next ship. They never came.[19] A poem by Francis Parsons, published in *Life* magazine in 1920, expressed the distress of American cavalrymen who went abroad without their mounts. The second and third stanzas of his poem "Regular Cavalry" read:

> They took as cavalry soldiers
> And made us machine-gun men.
> I said good-by to my old brown horse
> And turned him into the pen.
> I never knew what a friend he was,
> But something stuck in my throat—
> He wheeled—and stopped—and looked
> At me . . .
> Good Lord, that got my goat!
>
> It wasn't much of a picnic—
> That scrapping over there.
> There was times I thought I'd never last,
> And times I didn't care.
> I often thought of the horse I'd lost,
> And wished before I died
> I could go on him one last patrol
> Along by the Rio's side.[20]

Lessons about the future viability of the cavalry being used in its traditional roles (mounted and used in reconnoitering, raiding, and scouting) had to come almost entirely from examination of European cavalry experiences because American cavalrymen had gained little to no experience during the Great War.

Postwar Cavalry Response

Once the war ended, the American and British cavalry journals served as forums for cavalrymen to debate the cavalry's roles and value in modern war. In its first postwar issue, the American *Cavalry Journal* appealed to officers to submit examinations of the conflict that evaluated how cavalry roles had changed. It acknowledged that certain cavalry roles were no longer needed, observing that "the prominence of the new weapons and of the other services [had] . . . dimmed . . . [the] light" of the cavalry, but that the light was not yet extinguished.[21] The response was encouraging. Writers acknowledged that the relationship between aviation and the cavalry had changed, but there was still a need for mounted forces. The airplane would complement horse cavalry in the performance of tactical and strategic reconnaissance.

As before the war, cavalrymen assailed the tendency of Americans to accept too readily new technologies over old ones. The U.S. *Cavalry Journal*'s editor argued that military officers who believed the cavalry had no future because of new technologies were exchanging "apples for dead sea fruit." He thought that these officers were falling prey to the American national characteristic of "too ready adaptability" by swapping airplanes and tanks for cavalry before the new technologies had proved their abilities.[22] A 1921 editorial in the *Cavalry Journal* cautioned cavalrymen not to allow the "transitory predominance of gasoline and technical novelties" to convince them that they were "good for nothing."[23]

Major George S. Patton Jr., who would become famous as a general during World War II, admitted to his colleagues within the pages of the *Cavalry Journal* that before the war he had been guilty of being overenthusiastic about new weapons. He confessed that he was one of the many prophets for a new technologically motivated arm who based their arguments on little more than "oral proof." While he did not entirely admit that he and his contemporaries were wrong, Patton argued there were too many "thoughtless critics" of the cavalry and of other arms supposedly made obsolete by the late war. Patton maintained, however, that these technological innovations had an important place in the future.[24]

Warnings against overconfidence in the ability of new technologies to replace the cavalry did not emanate only from active cavalry officers. The *Christian Science Monitor* observed in 1918 that the war "taught us that prophecy . . . is futile," noting that the cavalry successfully drove the Germans out of St. Mihiel despite predictions to the contrary.[25] General John

Pershing claimed that while some "unthinking" people might exploit the minimal use of the cavalry in the war to argue its day had passed, he believed that even the cavalry's brief employment proved that it would continue to be an important military branch.[26] Major William C. Sherman of the Air Service also noted the "regrettable tendency on the part of the overhasty to assert that aircraft have rendered cavalry useless for future wars."[27]

Even before the cavalry journals resumed publication following the war, some aviators debunked the exaggerated claims of aviation capabilities. A member of Chief Aviation Officer General W. L. Kenly's staff reported Kenly's statement to a journalist that the "Air Service has already gotten into trouble with the American public by prophesying a performance and then not performing it."[28] A November 1919 Air Service report blamed those who were "ignorant of the capacity of an air service," including politicians and journalists, for overblown predictions about aviation's abilities. The report, without providing specifics, argued that American aviators "had been given an impossible standard to live up to" and that the American people and even the army "expected too great things of the Air Service."[29]

On the other side of the Atlantic, the British followed a similar path. Field-Marshal Earl Haig called for the British *Cavalry Journal* to record cavalry history and lessons from the war to "correlate in the light of the experience of the war the policy and principles of the training of cavalry and allied arms." Haig argued that the war had expanded cavalry duties and "shown them to be much more diverse and complicated than heretofore supposed." He believed that the *Cavalry Journal* should "act as a valuable guide to the mass of new matter that will gradually become available" and that technical articles should amplify and exemplify official training manual teachings.[30]

A 1927 assessment of British Army training also preached caution about placing too much confidence in new technologies. The Memorandum on Army Training warned that an open mind was necessary for progress but so was a need to cultivate "a sense of proportion" because "exaggerated enthusiasm for a weapon, blind adherence to some form of tactical action, an inclination to enshroud the skill of an arm in mystery, are not conducive to progress." The memorandum added that the cavalry had "made genuine and successful effort to adapt their training to new conditions."[31]

Yet no one could doubt that aviation technology significantly advanced during the four years of conflict in Europe (as appendix A illustrates), greatly increasing the capabilities of air units. The flimsy experimental aircraft in service in 1914, machines that had frequently killed their pilots, had disappeared. By the end of the war, specialized observation, fighter, and bomber airplanes had replaced the generic types available when the war began. Postwar aircraft were more reliable, flew faster and further, and had larger carrying capacities than their wartime predecessors due to improvements in aerodynamics and engine design.[32] Prewar airplanes had top speeds in the range of seventy miles per hour. The fastest postwar aircraft, fighter planes, reached speeds of 120 miles per hour in 1918 and almost 200 miles per hour by the end of the 1920s. The maximum speeds of reconnaissance and bomber airplanes also increased over the decade from about 100 miles per hour to 140 miles per hour. Of course, speed is less important than stability and carrying capacity for reconnaissance and bomber aircraft; these characteristics improved relatively slowly during and after the war.[33] Communications between airplanes and the ground also improved with the development of lightweight wireless radio sets for aircraft; aerial photography revolutionized the collection of information about terrain and enemy positions.[34] In light of these developments, cavalrymen could no longer dismiss airplane technology as low-powered, unreliable, or incapable of contributing to successful campaigns.

Arguments for the continuation of the cavalry despite improvements in aviation technology in the United States came from the *Cavalry Journal* as before the war and from the work of the newly established chief of cavalry. The United States Cavalry Association's journal no longer had to fill the place of a branch head in discussions of doctrine and military developments. This long-awaited leader was the result of the National Defense Act of 1920, which established chiefs of each of the combat services.[35] The *Cavalry Journal* rejoiced at the announcement of the new position and claimed it as a "cause for thanksgiving and a restorer of morale" because the holder of the post would assist the cavalry in future development.[36] Major Willard Ames Holbrook, an 1885 West Point graduate and commander of the army's Southern Department in 1918, became the first chief of cavalry, serving from the passage of the new law until his retirement from the army in July 1924. Like many of his cavalry contemporaries, Holbrook had been preparing for overseas duty when the armistice began and did not serve in Europe during the war. Instead, he commanded troops along the U.S.–Mexico border.[37]

Holbrook did not specifically mention aviation in his first article in the *Cavalry Journal*, but he stressed the importance of a spirit of cooperation and "team-play" between the different elements of the army. He maintained that the cavalry remained an "essential part of a well-organized army," a belief, he observed, shared by Great War leaders and the AEF Cavalry Board.[38] Holbrook stressed the need for the cavalry to prepare itself to confront any emergency by maintaining its mobility, firepower, and spirit of cooperation.

How, why, and with whom the cavalry could cooperate and why it was still needed was identified in the American and British cavalry journals and in the *Employment of Cavalry, 1924–1925*, a work prepared under the direction of Holbrook. It represented the most thorough official thinking about the role and utility of the cavalry and enumerated the arguments American cavalrymen utilized to defend their value in reference to aviation. This Cavalry School document encompassed most of the postwar justifications explaining the continued necessity for cavalry in situations where aviation failed. Its primary focus was on reconnaissance, and it identified situations when the cavalry would still be required, most significantly in certain meteorological and terrain conditions.[39]

Although postwar aircraft could fly faster, travel longer, carry larger loads, and maneuver more easily than previous craft, *Employment* noted that they were still incapable of operating in certain weather conditions.[40] This argument also appeared in the American and British *Cavalry Journal*. Only two years after the war ended, Major General Leonard Wood, the former commander of the Rough Riders and chief of staff, argued that the cavalry would remain necessary for reconnaissance in weather conditions that grounded aircraft.[41] A British major in the Royal Engineers, the unit originally responsible for military aviation in Britain, noted that aerial observation was only possible in "reasonably fine weather."[42] A major and instructor in tactics at the Cavalry School warned that there would be many times when fog and rain would prevent airplanes from collecting necessary information.[43] Citing British maneuvers and French operations in Africa in 1929, a captain in the cavalry reserve observed that "bird-men stay on the ground" when it rained, meaning that "the day of cavalry is not yet done."[44]

Although flying might be possible at night, *Employment* argued that aerial observation was not.[45] Decreased visibility made observation difficult and landing dangerous. An American aviator stationed on the southern border argued that it was especially hard to land a DH4 (the standard American bomber) in the dark because of the aircraft's design.[46] A British major concurred with his American and British counterparts that airplanes were more limited than "generally realized," as they could only be used during the day.[47] A 1921 U.S. *Cavalry Journal* article recounted a story that during the war, pilots attempting a night reconnaissance reported seeing a large body of Turks. They estimated that the force numbered from six to eight thousand troops and were marching toward the army's right flank. The sun revealed the "Turks" were actually "several enormous herds of sheep."[48] The incident highlighted the difficulty of identifying units on the ground from the air in darkness. How could nocturnal aerial reconnaissance be trusted if its practitioners could not even tell the difference between men and sheep?

"Lest Ye Forget," a poem in the British *Cavalry Journal* in 1922, also cautioned against depending on aviation alone for reconnaissance in poor weather and at night. This poem summarized the major arguments of both cavalry's detractors and supporters. In the first stanza, airplanes were described as a means of replacing the cavalry's reconnaissance role, yet in later verses the poem outlines the limitations of aircraft. There was danger at night in stormy skies if a nation depended only on aviation to seek out the enemy. The first stanza read:

> 'Tis said the cavalry is dead;
> That whirring planes high over head
> Will information gain;
> So why should cavalry maintain
> A place in modern wars?
> 'Tis said the cavalry is dead; that mars
> Needs not this mobile arm
> To guard against alarm
> When night or stormy skies
> Aid enemy's surprise.[49]

In a 1925 British *Cavalry Journal* article, a colonel made the argument simple saying, "Night comes and air information closes down."[50]

Terrain, *Employment* stated, could also limit aviation's observation capabilities. Locating, identifying, and counting enemy forces from the air was difficult when those forces were in terrain that provided easy concealment.[51] Major General Wood observed that the difficulty of aerial reconnaissance in terrain such as wooded country challenged the idea that the cavalry could be replaced entirely.[52] Aviator Stacy Hinkle observed that when attempting to locate two flyers who had crash landed in Mexico, one search airplane passed over them without its crew spotting them because of the brush surrounding the men on the ground. He also argued that the construction of the DH4, in use in 1919 on the border, made landing on rough ground impractical.[53] It is easy to forget that terrain affects the location of airfields as well as hinders observation. Aircraft require large, flat, and smooth areas for takeoffs and landings.

Cavalry Colonel Hamilton Hawkins claimed that reconnaissance was still "one of the important duties of the cavalry." He argued that should another trench war develop—which he thought unlikely—only then could airplanes "relieve the cavalry" of this important mission.[54] Even if trenches developed in future conflicts, a British poet claimed that the cavalry could capture them, citing an example from the war. He wrote:

> 'Tis said the cavalry's dead;
> Such is the word which not is spread;
> Yet pause to give a thought
> To lesson which Beersheba taught;
> Swift horsemen crossed that plain
> 'tis aid the cavalry is dead; how vain;
> When cavalry at work
> Took trenches of the Turk,
> Thus causing deep chagrin
> To infantry therein.[55]

In addition to cavalrymen, some airmen also thought the airplane was not developed enough to completely take over reconnaissance tasks, especially in the same situations listed by cavalrymen, such as rough terrain and when fog and mist reduced visibility, making flying difficult or impossible.[56]

Even when weather, terrain, and time of day were ideal for aircraft, American and British cavalrymen argued throughout the 1920s that aerial reconnaissance could not collect all of the information mounted units could. Aircraft could not correctly identify ground forces or determine enemy intentions.[57] They also could not capture prisoners for information or determine the accurate strength of troops in wooded or mountainous country.[58] In 1919, an American pilot argued that operations on the border required the cavalry to confirm who had been shot and if they had died after aircraft had fired at mounted men on the ground.[59] In the mid-1920s, one British colonel maintained that even in ideal flying conditions, the RAF could not obtain identifications, which he called "priceless little items of information which enables the general staff to piece together the enemy's plan."[60] Late in the decade, cavalrymen continued to maintain that aircraft lacked the ability to "distinguish easily friend from foe."[61] The air service could not accurately secure key data such as identification, strength, composition of units, or, most importantly, their intentions.[62] As the 1923 U.S. *Field Service Regulations* noted, these intentions could be acquired only through the capture and interrogation of prisoners, seizing posts and telegraph offices, and examining letters, dispatches, and newspapers, all activities that aircraft simply could not do.[63] This lack of flexibility in gathering battlefield intelligence supported the continued existence of and cooperation with the cavalry.[64] RAF Wing-Commander T. L. Leigh-Mallory concurred with the assessment of aviation's limitation and consequently supported combined action between the cavalry and the RAF in reconnaissance operations.[65]

Cavalrymen also declared that negative information (reports that no enemy troops were found in the area under investigation) gathered from the air could not be trusted as much as information gathered from the ground.[66] *Employment of Cavalry* contended "cavalry, but not the air service, can obtain complete negative information."[67] Reliable negative information from the air was limited because of terrain and the operational characteristics of airplanes. They could not search woods or inside buildings, nor could they hover over locations. Not only was negative information from the air of little value, according to a cavalryman, it could be "extremely misleading."[68] An 8th Hussar major declared "if aircraft report that they have flown over a locality—village or wood and have seen *no* signs of the enemy it does *not*, in the least, follow that *no* enemy is present in these places."[69] The 1st Surveillance Group, operating as part of the Army Air Service border patrol on the U.S.–Mexico border, discovered the cavalry had to locate the small groups that hid under cover when airplanes passed overhead.[70]

Cavalrymen also defended the continued value of the cavalry in reconnaissance by arguing that no matter the conditions on the ground or in the air, airplanes could not maintain constant contact with and observation of the enemy. In December 1920, the U.S. War Department's *Cavalry Memorandum No. 1* stated that the cavalry could maintain constant observation, but aerial reconnaissance was "necessarily intermittent."[71] This U.S. Army Command and General Staff College publication observed that airplanes had to land to refuel and replace worn-out parts and pilots. Even if multiple aircraft provided continuous aerial observation, the information gathered by the first pilot could not be transmitted to his relief because the second pilot would be in the air before the first pilot landed and delivered his report (this problem was more acute before the development of small radio sets for aircraft).[72] Even at the end of the decade, a 1929 UK *Cavalry Journal* article noted that airplanes could not maintain continuous contact with enemy forces, unlike the cavalry that could stay in close visual or direct contact with opposing forces.[73]

Another factor militating against the exclusive use of aircraft in reconnaissance was their dependence on landing fields and facilities near front lines.[74] As one airman stated in 1920, the best method to deliver reconnaissance reports and get orders from commanders was to land and receive instructions verbally, but the speed of modern airplanes made it "impracticable for them to land without 'crashing,' save in fairly large and unobstructed fields; too often such are unavailable."[75] The concern was not just theoretical. Hinkle argued that the lack of landing fields in the American south hindered the air service border patrol's ability to perform its duty to "protect the entire length of the American side of the United States–Mexico International Border."[76] Although cavalrymen recognized that additional and better-placed landing fields could remove this limitation in the future, they used this factor to demonstrate that they were still, for the time being, needed to maintain contact with headquarters.

Juvenile, Emotional, and Irrational Arguments

While many arguments for the continued existence of the cavalry were carefully reasoned, based logically on the technological and operational limitations of aircraft at the time, a few contentions appeared less well-grounded in facts. Not all cavalrymen in the 1920s followed the lead of the Cavalry Board, the chief of cavalry, and the *Employment of Cavalry* by rationally assessing the abilities of the cavalry and aviation. Some of the cavalry's defenders resorted to personal attacks against those perceived to be cavalry detractors and contended that the horse and man were more valuable above all else in war.

Name-calling and questioning their opponents' intelligence were among the more childish of these approaches in both the United States and Great Britain. Those claiming the days of the cavalry were over were labeled "doubting Thomases" and "armchair doctrinaires" who possess questionable intelligence and knowledge.[77] British General Lord Horne argued that those calling the cavalry obsolete were unlikely to be experienced soldiers or men who had "responsibility of high command in war." Horne believed anyone who argued for the cavalry's abolishment had misread the lessons of the war, particularly its first weeks and operations in Palestine.[78] A U.S. Army captain noted that "the present rage . . . over 'mechanization' and 'motorization'" in the British Army was "an exaggerated expression of flighty opinions held by people whose recent vision is restricted to the Western front." In contrast, "sober military judgment" recognized the "future utility for cavalry" in countries similar to Palestine.[79] The British *Cavalry Journal* reported that both in the dominion and abroad, magazines called the cavalry's opponents "thoughtless critics" and "tank-obsessed, Ichabod-calling, mechanicalised Jeremiahs."[80] A British *Cavalry Journal* article summarized the feelings toward those who said the cavalry was obsolete by stating simply, "You don't know what you're talking about."[81]

One theme popular with cavalry of all ranks was faith in the man (sometimes expressed as man and horse) over the machine, the contrast of the moral superiority of living beings to cold, unfeeling machines. This motif of man overcoming all technological challenges was rare in prewar military writings but blossomed during the 1920s.[82] First Lieutenant Fenton Jacobs mastered the use of vivid descriptions to engage his readers' emotions, writing in 1921 that "whenever and wherever the lure of the open appeals and red blood surges in the veins of those who love adventure, dash, and romance, there one will find the mounted man in his glory and predominance," a warrior free to toss his "cigarette into the air and unrestrained, launch himself right at the foe." Pilots, however, "must hold hard, cool, and deliberate to launch their projectiles with mathematical precision along trajectories through miles of space."[83]

The belief in the soldier's superior fighting power was not confined to a few people or limited to the cavalry. Although few writers could compete with Jacobs's flowery, romantic descriptions, other cavalrymen compared men and machines. Lieutenant-General Sir Alexander Godley argued at the Royal United Service Institute in January 1922 that "whatever modern inventions and mechanical appliances there may be, you may always, in the end, have to fall back on the combination of the man and the horse."[84] General Lord Horne added that "clear thinking will lead us to decide that the day is not yet when mechanical and other contrivances can take the place in war of either the man with the rifle, or the man on the horse."[85]

Field Marshal Earl Haig clearly displayed his sympathy for the human element when he proclaimed in 1922, "I certainly am not among those who hold that cavalry is a dead arm, or that the place of flesh and blood, in man and horse, can ever be wholly taken by petrol and machinery."[86] Three years later, Haig responded to accusations that the military horse was obsolete by arguing that airplanes and tanks were "only accessories to the man and the horse" and that the horse would have as much a place in the future as in the past. New inventions, he declared, "always produce an antidote."[87] One American cavalry colonel summarized the argument succinctly: "The fighting man, not the fighting machine, will always continue to be the principal means of making warfare."[88] The report on the 1927 British collective training period stated that "in future, as in the past, the trained man, whether commander or soldier, will be the chief factor."[89] Brigadier General James Parker concisely stated the feelings of many cavalrymen asserting "the idea of displacing cavalry with machines is preposterous."[90] As late as 1929, Major K. S. Bradford argued the cavalryman was "nimbler than any machine. He is silent as well as swift" and that "the real sinew of war is not wealth, but man—and the horse."[91]

This common theme of men's superiority to machines harkened back to the old idea of a "cavalry spirit." Intangible and unmeasurable, this mystic force made success possible in the most hopeless situations. One cavalryman argued that this spirit "often gave life and soul to an army which otherwise were a dead machine in some countries."[92] Liddell Hart called cavalry spirit "the very soul of war."[93] This element, unique to cavalry, was viewed by those both inside and outside the cavalry as something to be cherished and maintained even if the cavalry itself was eliminated. In 1927, Haig warned the Sub-committee on the Strength and Organisation of the Cavalry that once lost, the cavalry spirit could not be reproduced in a minute.[94] The idea of transferring the spirit to other units, thus preserving it, gained many supporters as it became increasingly clear that tanks might replace horses. Winston Churchill, an advocate of the continued reduction of cavalry in the postwar period, argued that he was neither against the cavalry spirit nor its qualities but both "ought to be married to any mechanical mobile development."[95]

Building a Cooperative Relationship

Although cavalrymen continually listed aviation's limitations, many recognized that their services' continued usefulness required cooperation between the air services and the cavalry. The various and indisputable shortcomings of aerial reconnaissance provided cavalrymen with opportunities to present their case for cooperation and coordination. The editor of the American *Cavalry Journal* encouraged readers to be "ready-tongued" to point out the fallacies of those who may claim that reconnaissance could be done by a single service or technology.[96]

Yet which service had the primary responsibility for reconnaissance was not yet clear. Mirroring AEF Superior Board and Cavalry Board conclusions from 1919—which themselves reflected prewar arguments—cavalrymen pointed out that airplanes "supplement and extend, they do not replace."[97] But as an American cavalry captain argued, aerial reconnaissance was of the greatest value, but to be effective it "must be supplemented by the work of ground troops," an idea codified shortly thereafter in the 1923 U.S. *Field Service Regulations.*[98]

A similar debate took place across the Atlantic. A British *Cavalry Journal* article reflected upon the relationship between aviation and the cavalry in a less technical and more literary manner. Written partially from the perspective of birds, this piece emphasized the military importance of having both land and air forces.[99] The father bird explained to his son why they could not avenge the murder of a fellow bird: "On the ground, and in the ground, lie all sources of life. We birds lost all hope of mastery of the ground when we took to the air. The fishes lost all hope of mastery of the ground when they took to the water, but man was wiser. He put mastery of the ground first, and now he is rapidly becoming master of air and water as well."[100] In the second part of the article an old cabinet minister tells a young air marshal that "only by the closest combination of all our forces—land, sea, and air . . . can we hope to be victorious over our enemies."[101]

No matter which service supported the other, the need for cooperation was clear. Numerous cavalry journal articles in both the United States and Great Britain stressed the importance of coordinating activities of airplanes and cavalry.[102] Each had virtues that compensated for the other's liabilities. As before the war, American and British cavalrymen proposed that aviation was not an enemy of the cavalry but a technology that strengthened mounted units. An American first lieutenant in the 6th Cavalry argued that airplanes, along with other new inventions, such as tanks and armored cars, relieved the cavalry of some reconnaissance work "so that men and horses" were not "expended unnecessarily" and could thus be fresh for use on the battlefield.

He believed the cavalry benefited more from aviation than any other branch because cavalry so often operated in close proximity to the enemy.[103] Edmund Allenby, commander of cavalry on the western front during the Great War, also argued that airplanes strengthened and increased the cavalry's mobility and augmented its "battle value" because of the greater field of vision possible from the air.[104] British Lieutenant-General Sir Philip Chetwode argued similarly that "modern inventions will not displace mounted troops" but that mounted troops could become "very much more powerful" by using such technology as airplanes.[105] Their postwar argument echoed prewar thinking that new technologies facilitate but do not replace the cavalry.[106]

To understand how best to take advantage of aviation, cavalrymen examined the recent war to identify the most effective techniques for air–ground cooperation. Campaigns in Mesopotamia and Egypt in 1918 demonstrated the "excellent work" done by the air force, cavalry, and infantry in concert, as testified to by British Lieutenant-General H. D. Fanshane and Lieutenant Rex Osborne of the hussars.[107] Most of the reports focused on communication between aircraft and mounted forces and identified both successes and failures.

Ground troopers seemed pleased by the way friendly airplanes could aid communications. A cavalry lieutenant argued in a 1920 article in the U.S. *Cavalry Journal* that airplanes equipped with radio sets could increase the radius of action for cavalry units by relaying information and instructions between cavalry in the field and commanders in the rear.[108] In one of the more than a dozen articles examining the British cavalry campaign in Palestine, Colonel George Mitchell of the General Staff Corps published details about reconnaissance reports dropped by airplanes, which assisted the cavalry considerably in its movements.[109] Captain J. R. H. Cruikshank, a subaltern under Allenby, noted how on a sunny cloudless day airplanes could receive messages from sheets laid on the ground in particular patterns and then reply by Morse code.[110] This system helped hasten the transfer of information during the campaign in Palestine. The advantage was that the airmen did not have to land to relay their messages, a significant problem in the days before airborne radio became common. Although attempts to establish rapid and reliable communication between airplane scouts and cavalry patrols were not always successful, British cavalry operations in the Middle East suggested the possible rewards of close cooperation in the form of rapid and direct communication, according to two American observers attached to British units.[111]

Writers had difficulty identifying similar situations when American cavalry troops worked with aviation. The 1918 St. Mihiel offensive was one of the few occasions when the United States Cavalry actually participated in its traditional remount, liaison, and patrol duties. However, it did not have much opportunity to work with airplanes. The American cavalry's wartime experience with aviation, according to a captain of the 2nd Cavalry, consisted mostly of avoiding aerial observation by enemy aircraft and wishing the American air service was as active as the enemy's.[112] Another cavalryman related that troopers were employed "extinguishing fires and rescuing French civilians from their wrecked homes" after bombs were dropped from airplanes.[113]

The shortage of examples of direct cavalry and airplane cooperation did not discourage postwar efforts for additional collaboration but rather stimulated them. General John J. Pershing, AEF commander, encouraged the continued development of cavalry cooperation in his message to the cavalry in April 1920.[114] One British officer complained that the war provided too few opportunities for the cavalry and aviation to work together, even though they were "so well adapted for mutual support."[115] Instead of setting airplanes and cavalry in opposition, many postwar articles by cavalrymen supported a necessary and positive interdependence, arguing for more opportunities for horse cavalry to work side by side with aviation.[116]

Postwar articles demonstrated a demand among cavalrymen for training and maneuvers with the air service to improve cooperation and coordination between the two branches.[117] One such article called for additional opportunities to work together to eliminate the possible ignorance and prejudice about the role and capabilities of aviation that, the author believed, may have hindered successful cooperation during the war. A. W. H. James, author of a series on aircraft and cavalry cooperation published in the British *Cavalry Journal,* supported requiring horsemen to learn about the actual capabilities of aircraft and focused on cooperation.[118] Cavalryman Major LeRoy Eltinge, formerly brigadier general and deputy chief of staff of the AEF, recommended that "every opportunity should be taken to hold maneuvers in conjunction with large bodies of all arms."[119] The solution seemed clear: more cooperative training focusing on learning "the fundamental principles of the other" could only improve overall efficiency.[120] For the cavalry and aviation to perform their duties optimally, each branch needed to understand the limitations and capabilities of the other, an understanding that could only be formed through frequent communication as well as combined training and maneuvers.[121]

American Operations and Maneuvers

American cavalrymen experienced their first postwar practice with aerial cooperation during operations on the U.S.–Mexico border. The United States organized patrols of its southern border to defend against hostile bands (some paramilitary, others bandits taking advantage of the instability) roving on the frontier. In June 1919, Army Air Service Chief Major General Charles T. Menoher ordered eighteen DH4 airplanes—twelve from the 20th Aero Squadron and six from the 11th Aero Squadron—to patrol from San Diego, California, to Brownsville, Texas.[122] These bombers were British designed but built in the United States and powered by Liberty engines, designed and constructed in the United States. The DH4's speed and service ceiling were improvements on previous aircraft, but its unprotected gas tank, pressure feed system, and exposed gasoline line were vulnerable to damage from enemy fire. In addition, the location of the main fuel tanks between the pilot and observer complicated communications.[123]

In the Big Bend of Texas, cavalrymen of the 5th and 8th Regiments patrolled on the ground, and airmen of the two assigned squadrons flew circuits two or three times a day, watching the border for any disturbance.[124] The standing instructions for aviators were to search for bands of men and, if found, to determine their number, location, and direction of movement. Radios weighed too much for the DH4s, so reports in the form of handwritten messages and sketches had to be dropped at the nearest cavalry outpost for ground action.[125] These were placed in white canvas bags with red streamers for ease of visibility. Initially, this one-way communication was the only way airmen could contact cavalrymen until landing.

The Army Air Service border patrol and the cavalry began working together on a better way to communicate between air and ground units before the end of 1919. Within a few months of June 1919, five more air squadrons returned from Europe and joined the operations on the border. The new 1st Bombardment Group, comprised of the 11th, 20th, and 96th Bombardment Squadrons, 12th Observation Squadron, and 104th Surveillance Squadron, patrolled from Arizona to Texas.[126] By late October, they established a limited two-way communication system after ground forces created a panel system using white canvas strips to form predetermined messages and requests. Approximately two months later, an improved system utilizing signal flags increased the range of messages from the ground to the air. However, this system, according to a participant, remained unsatisfactory for aviators trying to read messages while battling wind turbulence and trying to write in an open cockpit while circling above a signaler.[127]

In addition to the routine operations, the air service and the cavalry had an opportunity to test their new procedures when one of the border patrol's airplanes disappeared somewhere over Mexico in 1921. The expedition to recover it teamed up the 12th Cavalry with aviators. Although communications between the forces proved far from perfect, this operation demonstrated that work should continue.[128] Pigeons successfully facilitated air-to-ground communication, but a lack of sufficient ground signal panels complicated information sharing. A cavalry pack train also assisted a liaison plane that went down early in the recovery operation.[129] These operations proved insufficient to formalize cooperation between air and ground units, making additional training required. Major William Sherman of the Air Service noted that air–ground communications had improved in these limited experiments but they remained deficient.[130]

By mid-1921, the Americans had accomplished their objective of ending raids and property damage by Mexican bandits and revolutionaries, so the Army Air Service border patrol was disbanded and its planes were reassigned.[131] Although short-lived, the patrols provided an opportunity to test and apply techniques of cooperation between air and ground forces. Hinkle argued that the experiences along the border demonstrated the need for better two-way communication and the importance of proper liaison.[132]

The American cavalry's endeavors to foster and perfect cooperation with aviation did not end after the border patrol was discontinued. Throughout the 1920s, cavalrymen participated in joint training and maneuvers with attached air units to satisfy their desire to improve understanding between the cavalry and the air corps. The training exercise at Camp Meade in 1922 allowed the 62d Cavalry Division to participate in a reproduction of a French battlefield, replete with "real ammunition, infantry, machine-guns, artillery, tanks, trench mortars, bomb-throwers, smoke screens, gas and air service." Airplanes maintained communication with ground signal units. The exercise spurred W. P. King to encourage cavalrymen to participate in similar future exercises.[133] The maneuvers of the 1st Cavalry Division in the fall of

1923 also provided opportunities for cavalrymen to work with aviation. Excellent weather and favorable terrain devoid of overhead cover allowed the army to gain experience in the use of signal panels with aviation. Contact planes also tracked the whereabouts and progress of various friendly and hostile units.[134]

Additional maneuvers throughout the 1920s, including the Air-Ground Maneuvers in San Antonio, Texas, May 15–21, 1927, provided opportunities for the cavalry and the air service to learn how advances in technology changed each other's strengths and weaknesses. The maneuvers in San Antonio simulated operation of a field army with an attached bombardment group. These maneuvers helped to develop teamwork between air and ground units and facilitated the creation of the fundamentals of air action. Major General Ernest Hinds reported that the limits of the air force were recognized as well as how powerful it could be as an auxiliary arm.[135] A cavalry major attached as a liaison officer reported to the head umpire that the maneuvers went "quite satisfactory" with excellent use of cover; however, there was only one opportunity to communicate with airplanes.[136] The press release of the maneuvers noted, "Absolutely no good can be accomplished by giving out any conclusion which belittle any branch whatsoever . . . every effort should be made to boost the interests of all branches."[137]

In late July 1927, a class from the Air Corps Advanced Flying School at Kelly Field arrived at Fort Bliss, Texas, to train with the 1st Cavalry Division. An article about the exercises in the *Cavalry Journal* described them as "very valuable" because they provided the cavalrymen "a clearer understanding of the powers and limitations of the air corps" and a "desire for further training."[138] The forty-five planes assigned to these joint training operations included one Douglas O-2 observation plane, one Douglas C-1 transport plane, fifteen AT-4 trainers (variants of the Curtiss Hawk fighter series), twenty-seven DH4s for attack or observation, and one ambulance plane of unidentified type.[139] Cavalrymen gained a clearer understanding of what aircraft could accomplish by watching and participating in dozens of demonstrations and operations with these airplanes. The most useful demonstrations, according to a participating cavalry observer, were tactical exercises testing cooperation with friendly airplanes, defense against enemy air attack, and communication between air and ground units. Major George Dillman of the 1st Cavalry reported that the observation planes provided ground commanders with "early and accurate information," demonstrating how aviation could "materially assist cavalry in rapidly moving situations." Dillman declared the exercise "very beneficial" overall.[140]

Additional maneuvers held in Marfa, Texas, in September 1927 involved a cavalry division with an observation squadron attached. According to Dillman, the air unit, the 12th Observation Squadron, functioned "in an excellent manner," but there were still some issues to work out.[141] Attack planes were also tested to gauge their effectiveness for use with the cavalry in offensive actions. Although the planes worked well, problems remained with cooperation. Dillman stressed the importance of frequent divisional maneuvers following garrison training to improve cooperation.[142] Similar tests at Fort Leonard Wood attached air units to ground units so they could learn about one another.[143]

The Army Air Service reciprocated the cavalry's attempts at cooperation. In his 1922 annual report, the chief of the air service noted the importance of combined training with field maneuvers in tactical problems at the various special service schools. Far from rejecting the cavalry, he stated the need to understand "one of the most essential features of peace time training," which was the "inter-relation between the air service and other arms."[144] Communication between the ground and the air remained a problem throughout the 1920s. American airmen rode horses to reduce the communications gap between the two services. An air unit's liaison officer had the unenviable job of accompanying the cavalry on training marches and maneuvers. One pilot assigned to such a duty described riding as "torture" because of the difficulty of getting the horse to follow directions and the "hot desert sun and the choking clouds of dust thrown up by trotting or galloping columns of horses."[145]

Combined training in the United States led some cavalry officers to consider the practice of temporarily attaching air units to cavalry divisions for training as insufficient. A unit outside a commander's peacetime authority might make him unlikely to use it during war. The Superior Board's 1919 report recommended that "a command and reconnaissance squadron [of] eight to ten airplanes be made an integral part of the cavalry division."[146] Cavalrymen also proposed including airplanes within cavalry units or at least attaching them to cavalry units during appropriate operations.[147] While an American officer attached to the Desert Mounted Corps headquarters in Palestine was pleased by the dropped messages that provided the results of morning aerial reconnaissance and the occasional supplemental notes, he regretted the absence of airplanes attached directly to cavalry headquarters. He suggested that airplanes be directly employed by mounted units as well as the infantry corps and army headquarters.[148]

Events and discussions in other countries supported the cavalry's desire to cooperate with aviation. American observers with the wartime British cavalry had suggested how this kind of collaboration could work.[149] A 1922 French conference of general officers and department heads garnered attention from American cavalrymen who believed the conference provided valuable ideas about how to combine the cavalry and aviation. The conference showed that "the history of the employment of the air service indicates a necessity for the closest co-operation and team-work between the air service unit and the ground troops. This co-operation cannot be acquired successfully after the initiation of hostilities, bearing in mind also that the cavalry will be active from the first, and must be insured by the inclusion of air service units as part of the cavalry divisions for their peace-time training." The French conference concluded that every cavalry corps should have a squadron of airplanes attached, with the air service and ground troops closely cooperating.[150]

Five years later, the American chief of staff decided to "incorporate in each cavalry division an observation squadron, air corps" as part of the War Department's program to increase the cavalry's future efficiency and usefulness.[151] His subordinates agreed that aircraft added to the cavalry increased the "battle value of cavalry."[152] Major General Herbert B. Crosby (1926–30), the third chief of the cavalry, argued that "an observation squadron of airplanes should be an integral part of a cavalry division" after reviewing the experiences of the 1st Cavalry Division in Fort Bliss, Texas, in 1928.[153] Crosby reported that the observation squadron would become effective if the air corps could "supply suitable equipment and the necessary personnel."[154] Aviation supporters opposed such plans because they hoped for an independent aviation branch, not the subordination of air units to ground units.[155]

British Maneuvers

Air cooperation in Great Britain with the cavalry was more problematic than in the United States. British airmen increasingly shunned cooperative roles, preferring independent action, both to justify their new doctrine and to save as much money as possible, as will be explored in later chapters. British doctrine asserted the importance of combined forces, yet conflict between various military units emerged as a result of the organizational structure of the British armed forces. The RAF was a separate service from the army, made independent on April 1, 1918, partly as the result of wartime expediency. Some British military personnel and government officials believed that combining the Royal Naval Air Service (RNAS) and the Royal Air Corps (RAC) into the RAF would prevent duplication and competition between these branches, providing better aerial protection of the homeland.[156] Thus in Britain, the cavalry did not work with the air force as a coequal branch of the same service, as in the United States. It had to learn to operate separate if roughly parallel administrations. This made combined training more difficult to coordinate because of the additional administrative steps required to arrange such exercises. However, British combined training was more structured than its American counterpart. The British divided training into an individual training season and a collective training period each year to train different combinations of units. These training exercises began in 1923 and continued into the 1930s.

Although early British maneuvers were limited in their very nature, they provided concrete information about the capabilities and limitations of the cavalry and the air services when attempting cooperation. These operations supported the conclusion that neither the cavalry nor aviation could fulfill the other's roles. A report prepared after the 1925 British collective training period warned that reconnaissance purely from the air was often flawed, so ground "commanders must realise that . . . tactical reconnaissance air reports, though valuable, are by no means infallible. Undue reliance must not be placed on these reports, and, where possible, air information should invariably be checked by ground reconnaissance."[157] Training reviews from 1926 further warned about the need to maintain cavalry for reconnaissance purposes because weather limited aviation's effectiveness. The weather that year was particularly bad, demonstrating the "necessity for reconnaissance by cavalry."[158] The report of the 1927 maneuvers also stated that the cavalry was needed to pierce the enemy's protective screens on the ground.[159]

Unlike in the previous training cycles, the 1927 training period gave the cavalry the opportunity to work with the RAF's Army co-operation squadrons. These units were nondivisional, as described in the *Memorandum on Training Carried Out during the Collective Training Period, 1927*; entire squadrons would be at a division's disposal during war. Despite the infrequency of army maneuvers and the few opportunities for training commanders in aerial reconnaissance, the report concluded that great advances occurred in cavalry brigades giving orders and instructions to the RAF as a result. A supplementary report concluded that the cavalry remained necessary because of the many shortcomings of aerial reconnaissance—the same ones that cavalrymen had been identifying for years in the cavalry journals—such as limitations at certain heights due to weather, gaps in coverage, and communication difficulties. The memorandum suggested that the cavalry and air units continue to work on communication during next year's training period.[160]

Additional maneuvers, including the Cavalry Staff Exercise in Oxford, gave cavalrymen more opportunities to work on cooperating with the RAF in reconnaissance as well as practicing techniques to avoid observation by enemy aircraft. The Cavalry Staff Exercise confirmed that the RAF was best for conducting distant reconnaissance, armored cars were best for engaging in medium reconnaissance, and the cavalry was best for completing close reconnaissance. However, in bad weather or when roads were poor, the cavalry would assume responsibility for all reconnaissance activities. In addition, the report of the exercise argued that even in cases of good weather, the cavalry remained the only ones able to "particularise." Aerial reconnaissance could collect information of a general nature, but the cavalry provided the detail. In their attempts to avoid detection from the air, the report concluded that the cavalry could easily disappear from view. The report agreed with previous training reports that communication between aircraft and the cavalry needed improvement.[161]

Cavalrymen may have been willing to cooperate with the RAF, but airmen were less concerned about working with the cavalry than with preserving their independence. Reductions in military spending in Britain in the 1920s caused many to believe that it would be operationally efficient (and more economical) to disband the RAF and return its component units to the army and navy. The Army Council, the governing board of the British military, stated in 1925 that coordination was easier between two services rather than three, suggesting "the desirability of restoring to the Army its military air arm."[162] Sir Hugh Trenchard, the first (January 1918–April 1918) and third (March 1919–January 1930) chief of air staff, avoided the need to defend the benefits of three services cooperating instead of just two by emphasizing the RAF's independent roles and downplaying its cooperative ones. He was supported in this effort by Winston Churchill, successively secretary for war and for the colonies between 1919 and 1922. During the years of debates over reducing expenditure, Trenchard claimed that the RAF could not and should not be diminished. Although he was willing to concede that aerial support of troops on the ground was beneficial, for Trenchard its importance paled in comparison to independent missions. He argued that if any reductions were made, they should come from the "incidental role" of assisting the army and the cavalry, not from strategic bombing.[163] Trenchard testified that the RAF could accomplish various reconnaissance and protection services with or without cavalry cooperation.[164] These contentious debates over the funding and organization of the British military did not provide an environment favorable to the development of close cavalry–RAF cooperation.

Another Challenge from Aviation

Although cavalrymen provided numerous reasons why aviation could not replace their branch in reconnaissance, they had no good defense for the horse's vulnerability to aerial attack. They recognized the necessity of defending their units from bombing and strafing, which placed a premium on not being spotted by enemy reconnaissance aircraft, as George Mitchell, an American observer with the British cavalry under Allenby, noted.[165] Ground strafing by enemy aircraft was of "special interest" for the cavalry because the "vulnerability of the horses exposes cavalry especially to this menace."[166]

The cavalry responded with innovative training and tactics. Solutions included not smoking on night marches and keeping under the cover of woods when marching during the day.[167] It was increasingly clear to cavalrymen and other military officers that unless they gained air superiority, freedom from hostile aircraft was unlikely for large masses.[168] Major Patton suggested cavalry always keep mobile to prevent an aerial attack.[169] The threat was not just theoretical. According to an American observer with the British cavalry, a night cavalry march in Palestine during bright moonlight resulted in the death of many chargers when an enemy plane "swooped down out of the skies and machine-gunned the picket lines."[170] A report from the 1925 British collective training period observed how "it has now become second nature for troops to seek concealment from the air on all occasions."[171] Firing at enemy aircraft from the ground had not proved an effective deterrent.[172] At the end of the decade, British *Field Service Regulations* stated that the cavalry still remained "very vulnerable to attacks from the air."[173] The threat of bombing or strafing was of increasing concern to cavalrymen.[174]

Doctrine

Guidelines for the various types of reconnaissance slowly emerged in the decade after the Great War. Incomplete experimentation with and the technological limitations of aircraft had not produced a clear doctrine prior to the war. Aviation could be used in strategic reconnaissance to aid the cavalry, but other forms of reconnaissance had to be left primarily to ground forces. Postwar military doctrine established clearer demarcations between aviation's and the cavalry's reconnaissance duties. In the 1920s, studies and combined training exercises led to the creation of manuals, field service regulations, and military school handbooks that provided detailed explanations of the joint uses of aviation and the cavalry in communication, reconnaissance, and security.

The core lesson in both the United States and Britain was the importance of combined operations. The United States Army American Expeditionary *Report of the Superior Board on Organization and Tactics*, published in 1919, supported a close coordination of the cavalry with air forces, a position shared by the Cavalry Board. The 1923 U.S. Army *Field Service Regulations* repeatedly stressed that success in war could only occur when all branches and arms of service worked together.[175] Not only was cooperation necessary for success, "the special characteristics of each arm adapt it [the arm] to the performance of special functions in execution of the mission of the unit in which the action of all is combined."[176] The 1924 *Field Service Regulations* of the United Kingdom echoed this thought, stating "the full power of an army can be exerted only when all its parts act in close combination and this is not possible unless each arm understands the characteristics of the other arms. Each has its special characteristics and functions, and is dependent on the cooperation of the others."[177] For example, in the absence of cavalry and aircraft, "the other arms are hampered by ignorance of the enemy's movements, cannot move in security, and are unable to reap the fruits of victory."[178]

This postwar doctrine provided for the proper employment and levels of cooperation between the cavalry and the air forces for roles such as re-connaissance, security, and protection. First, both British and American doctrine divided reconnaissance into three types: distant (strategic), close (near-strategic), and battle (tactical). Cavalry and airplanes received specific assignments for different situations.

Distant reconnaissance—defined loosely as the collection of information about distant objectives for the creation of strategic and operational plans prior to, during, and following hostilities—was the duty of both the air service and the cavalry; however, aviation was, according to *Cavalry Training*, capable of "carrying out distant strategical reconnaissance far beyond the reach of mounted troops."[179] British doctrine considered RAF aircraft "specially [sic] suited for long-distance work" and distant reconnaissance would "normally be a duty of aircraft," but their reports "must be verified and amplified by the cavalry."[180] By the mid-1920s, military theorist Basil Henry Liddell Hart claimed, "By the universal consent of all general staffs, aircraft have replaced cavalry as the means of distant reconnaissance, leaving to cavalry the duty of close reconnaissance and acting as a protective screen within a short radius of the main forces."[181]

American policy was similar. The second-place winner in the U.S. *Cavalry Journal's* 1923 Prize Essay Contest included a chart of the most appropriate reconnaissance roles for airplanes and cavalry. The chart compared the abilities of the cavalry and aviation to complete various types of reconnaissance. The author, cavalryman Edward Fickett, created two detailed tables identifying the service best equipped to perform tactical and strategic reconnaissance missions.[182] Not surprisingly, aviation was considered best for strategic work, whereas cavalry was better suited for tactical operations. Fickett clearly concluded that each service required the other to assist in areas where it was less suited.

Close reconnaissance is the collection of information for tactical decisions when opposing armies were within a few days' march of each other, gathered by both aviation and the cavalry in the "service of large units and by each arm in connection with its own operations," as American army manuals noted.[183] In Great Britain this duty was performed by the Army co-operation squadrons, the division of the RAF that aided the cavalry. In tactical reconnaissance, as in strategic reconnaissance, information obtained by aircraft was still required to be "confirmed and supplemented by cavalry, or other [ground] troops."[184]

The last form of reconnaissance, tactical reconnaissance, was assigned to all combatant arms, including the air service and the cavalry. Clearly dividing the responsibilities between aviation and cavalry proved difficult. As Air Corps Major H. H. Arnold, commander of the 16th Observation Squadron at the Cavalry School, noted in 1928, "Their work interlocks and intertwines to such an extent that no exact line of demarcation can be drawn separating their fields of reconnaissance."[185] Consequently, the Tactical Principles and Decisions course on reconnaissance stressed that "in order to prevent duplication of effort, there is a constant interchange of information secured in close reconnaissance by the cavalry and the other arms."[186] Army regulations also discussed the necessity of close cooperation between the cavalry and the air services. Even with the use of both the air service and the cavalry, U.S. field service regulations stated that they were "not sufficient" to complete reconnaissance entirely. Other branches, including the infantry, might also be required to collect pertinent information for commanders.[187]

The cavalry and the air force also needed to cooperate in pursuit of routed units, distant action, and raids. British cavalry regulations in 1929 stated that pursuit of the enemy remained "one of the special duties of cavalry and aircraft working in co-operation." The regulations also argued that aircraft were essential in distant action and raids.[188]

In Britain, not surprisingly, cavalry policy supported cavalrymen's contentions that they were still needed and would need to work closely with aviation.[189] As the decade passed the cavalry's position began to lose support in military policy circles. The 1929 British *Field Service Regulations* altered the relationship between the cavalry and aircraft significantly. The broad term "mobile troops" replaced the word "cavalry" in descriptions of the relationship of providing security and winning the "fruits of victory." Cavalry was relegated to the end of the paragraph and stated to be "at a great disadvantage unless accompanied by artillery and armoured vehicles."[190] These changes demonstrated an adjustment in the value assigned to cavalry in relation to aviation and the rest of the army. Military policy no longer regarded the cavalry as an equal partner with aviation. It also reflected the changing relationship between the cavalry and mechanization. In 1928 and 1929, Britain converted the first two cavalry regiments into armored car regiments.

—

Utilizing experiences gained during the Great War as well as postwar maneuvers and exercises, cavalrymen continued to defend the utility of their branch in the 1920s. American and British cavalrymen continued to rely upon their prewar arguments while introducing new reasons why the cavalry had not become superfluous. Their older arguments included challenging the exaggerated claims of aviation supporters and listing airplanes' technological and operational limitations. Their newer arguments included discussions of landing field accessibility, the unreliability of negative information from airplanes, and the inability of aviation to maintain direct contact with the enemy.

As they had argued before the war, cavalrymen claimed that despite advancements in aviation, airplanes' limitations necessitated the cavalry's continuation. While the aircraft's technological improvements during the war reduced their operational limitations, cavalrymen argued that certain situations still required the cavalry. Not surprisingly, postwar discussions continued to center on reconnaissance, the role that most connected the two organizations.

Cavalrymen in the United States and Great Britain, supported by senior military officers, sought to apply the lessons of the Great War by actively cooperating with airplanes while fighting calls to eliminate their service. Cavalry journals, committee reports, and other publications as well as training, maneuvers, and doctrine reveal that cavalrymen were not rigidly opposed to aviation or aviators; in fact, they were actively searching for ways to work with aviation. Discussions of the new technology's limitations had less to do with a distrust or hatred of technology and more with a realistic appraisal of the airplane's abilities. The arguments for the cavalry's continued viability included more than the relative operational capabilities of horse cavalry and airplanes. As the next chapter demonstrates, economic and efficiency arguments proved more damning for the cavalry.

CHAPTER FOUR

National Economy
Aviation versus the Cavalry

While cavalrymen addressed the challenge of airplanes in classrooms, on training grounds and in the pages of journals, another set of battles occurred in capital buildings and the press, as political and military leaders battled over finances and the cavalry's fate. Here aviation scored a clear-cut victory over the cavalry by waving the sword of economy. Both the United States and Great Britain became enmeshed in national campaigns to reduce military spending after the huge expenditures of the Great War. Both countries had to address their security, both national (metropolitan) and imperial, in the decade after the war. In Great Britain and to a lesser extent in the United States, these sets of concerns overlapped. Government officials, aviation supporters, and military officers debated the relative value of the cavalry and aviation and their abilities to fulfill their commitments as economically as possible. Economy was attained partially by reducing the expenditure of all military branches, but the cavalry was the one branch continually singled out for reduction in direct response to the creation of the air force.

This chapter examines arguments promoting aviation at the expense of the horse cavalry utilized in Great Britain and the United States during the interwar period. These arguments appeared in political debates, popular and professional journals, personal correspondence, and official reports on national expenditure. Those campaigning to reduce the cavalry's size, role, and expense argued that the cavalry could not compete with new modern technologies on the battlefield, that it was too expensive to maintain, and that airplanes could perform national defense duties more cheaply.

This chapter also dissects two myths intertwined with the national expenditure and military reduction campaigns, particularly in Great Britain. The first myth was that air policing, the program for maintaining peace and stability through the use of airpower alone, was cheap and successful.[1] In reality, British use of air policing was neither successful without ground support nor did it prove to be as inexpensive as predicted. Yet the air advocates won the public relations battle; their assertions that air policing was cheaper than ground troops, such as cavalry, were accepted, and those who attempted to demonstrate the limitations of airplanes or defended the continued viability of horsed cavalry were overruled.

The second myth dealing with the stereotype of conservative and antitechnological horse cavalrymen is more difficult to outline. While the components of this stereotype are straightforward, it is unclear how widely it was held. The stereotype was that cavalrymen were opposed to any form of technological change—whether combat cars, tanks, motorcycles, or a combination of all of these—and unquestioningly against any reduction of horses in the branch. While there are countless examples of horse cavalry being called obsolete throughout the 1920s and 1930s, it is difficult to identify direct accusations that cavalrymen were antitechnological and reactionary. How much of the attack against the horse cavalry as an institution, or what Brigadier General Edward J. Stackpole Jr. called the horse cavalry's "obsolescence complex," was also applied to cavalrymen as a society is not clear.[2] Horse cavalry's supporters, however, certainly believed that the stereotype existed, as evidenced by their continued attempts to debunk it. Whether the stereotype was widely held by horse cavalry detractors or existed only in the minds of cavalrymen, it was not accurate. Many horse cavalry supporters actively campaigned for the horse cavalry's retention but only until an adequate technological replacement could be produced. Despite their willingness to transform once a reliable alternative was available, cavalrymen fought a battle against unnamed accusers that stereotyped them as backward. Although cavalry supporters were successful in keeping such critiques of cavalrymen out of committee reports, they were still not able to prevent the cavalry's reduction.

Because these topics are best described thematically, this chapter deviates from the loose chronological format of earlier chapters. Each individual subject is presented on its own time line, but the transition from one topic to the next requires a significant jump across some two decades' worth of debate and discourse, exploring the long-lasting economic and policy issues that characterized the interwar period. It also departs at times from an exclusive discussion of the cavalry, which is unavoidable due to the need to establish the larger context in which debates comparing the relative values of the cavalry and airplanes occurred.

Cavalry Stereotype?

As previous chapters have explored, cavalrymen had faced accusations that their branch was obsolete for decades, if not longer. Although they were ultimately unable to halt a reduction in cavalry numbers during the budget cutbacks of the post–Great War period, British cavalrymen and horse cavalry supporters utilized the forum provided by professional journals and postwar economic committee hearings to address what some perceived as a popular stereotype of backwardness. In the conclusion of his book *Doctrine and Reform in the British Cavalry, 1880–1918,* historian Stephen Badsey claimed that by the early 1920s the "cavalry and their generals became scapegoats for the perceived wider failings of the British Army on the Western Front."[3] In addition to claims that the cavalry was obsolete, Badsey described the development of the myth of the backward cavalryman tied to the past, unwilling to modernize. Once established, the myth of the old-fashioned, useless cavalry was impossible to shake, leading Badsey to call it a "zombie" myth, a story that once created will not die.[4]

While in most cases cavalrymen defended their branch from accusations of obsolescence, others responded to what they saw as personal attacks. While these cavalrymen almost universally failed to name their detractors, they argued that they were just as progressive as other military men. Major and Brevet Lieutenant-Colonel C. B. Dashwood Strettell stated in a 1921 lecture before the Royal United Service Institution that the

> discussion and argument as to the future role of cavalry which has so far taken place has, to my mind, been somewhat minimized in value by, on the one hand, the enthusiasm of the supporter of mechanical warfare leading him to somewhat didactically assume that cavalry officers are too conservative, and, indeed, too stupid, to move with the times, and, on the other hand, by a possibly righteous indignation on the part of cavalry officers at the assumption.[5]

Strettell recognized the need to adapt to modern mechanical devices and tactics while rejecting the contention that cavalrymen were opposed to change.[6]

Perhaps Major A. R. Mulliner of the 8th Hussars best described the feelings of his contemporaries when he wrote that "one feels somewhat like a barrister, who pleads in defence of a man charged with homicide against whom a mass of seemingly convincing circumstantial evidence has been collected and for whose conviction the populace clamours."[7] He clearly stated that cavalrymen did not "minimize the value" of airplanes or other new inventions. Instead, Mulliner argued that cavalrymen desired an impartial look at the limits of the new technologies proposed to fully replace the cavalry. He warned his government and fellow officers not to "be like the child that, given a new toy, in its delight at obtaining something new, throws away the old before unwrapping the parcel; then, when it finds that this new thing does not afford it the pleasure and amusement it formerly derived from the old, cries for its return, only to find that the shops have ceased to stock it."[8] Mulliner provided a clear path for the future of technological incorporation of which the cavalry would approve: "test, prove, and carefully unwrap this new thing and strip it of the glamour and attraction of the 'something new' that is such a dangerous fetish of the present day."[9] He expressed the desire for caution when evaluating new technologies, not outright and immediate rejection.

General Sir George Barrow echoed Mulliner's point, arguing that it was "rash as well as unscientific to make deductions from speculative imaginations instead of from observed facts and experiences."[10] He warned, "Do not let us be led astray, as has so often happened before by the verbose prophecies of those opponents of the cavalry arm who are but wise in their own conceit."[11] The cavalryman's question, another officer noted, was not "cavalry or machine" but "how to combine the essential characteristics of both."[12]

Horse cavalry supporters continually tried to refute the stereotype of backward, antitechnological cavalrymen in committee debates, particularly in Great Britain. In Great Britain, those who supported mounted troops were concerned about the widely held opinion that the cavalry was hostile to any change that would facilitate cavalry reductions or the branch's elimination. The records of the 1927–28 Sub-committee on the Strength and Organisation of the Cavalry reveal this anxiety. In his testimony before this body, General Sir Alexander Godley, former commander of the New Zealand Expeditionary Force and an organizer of irregular mounted regiments, declared that cavalrymen were actually "crying out" for machines. "Of course, they are," he observed, "they are not retrograde."[13] Another cavalry officer, Lieutenant-General Sir David Campbell, testified that "people imagine that there is a tremendous prejudice in the cavalry about mechanization, I do not think this really exists . . . the cavalry soldiers, when he sees that you are able to give him something that will enable him to carry out his work more efficiently, he will, and regiments will, accept it."[14]

Witness after witness encouraged the further incorporation of technology into the cavalry and stressed the cavalryman's willingness to receive it. One such witness, Colonel S. C. Peck, the War Office's director of mechanization, used his own experiences to defend the flexibility of horse cavalrymen. He described the successful mechanization of two field artillery brigades, one of which he described as having the reputation as "the most 'horsey' brigade of the Royal Regiment of Artillery." Although the brigades were initially resentful of the transformation, Peck noted that after only a short time, officers and men in both brigades "took the same pride and interest" in their new machines as they had in their horses.[15] He argued that cavalrymen would respond similarly.

Committee reports did not overtly use the cavalry stereotype to justify cavalry reductions. In fact, the Sub-committee on the Strength and Organisation of the Cavalry noted that any difficulty "anticipated in connection with the conversion of cavalry regiments into mechanized units" could be "treated as eliminated, since there is every reason to believe that cavalry officers . . . are quite ready to look facts in the face, and, with whatever natural regret, to take no exception to the changes from horses to machines."[16] Yet despite such support and the cavalry's efforts to work with new technologies and mechanization, the cavalry's detractors won the public relations and funding debate. The battle between the supporters and opponents of horse cavalry about mechanization closely mirrored the debates the advocates of horse cavalrymen and the proponents of aviation were experiencing with airplanes.

Great Britain—Air Organization

The conversations regarding the value of the cavalry and aviation occurred as Britain was trying to recover from the devastation of the Great War. The war cost the nation millions of people and pounds and eroded its political and economic dominance of the world. Yet after the Great War, the British had to find the manpower and funds to administer the territories they had acquired as a result of the liquation of the German and Ottoman Empires. Their military commitments included troops in India, Iraq, Palestine, Egypt, Constantinople, and the United Kingdom itself.[17] It was necessary for Britain to identify a means to fulfill its expanded obligations while simultaneously reducing its total military expenditure.

A national drive for economy consumed the British government after the conclusion of hostilities. In 1922, Henry Higgs, a historian of economic

TABLE 1

Selected Committees on National and Military Expenditure

	MEMBERS	MEETING/ REPORT DATES
Committee on National Expenditure (Geddes Committee)	Sir Eric Campbell Geddes (chairman), Lord Inchcape, Lord Faringdon, Sir Joseph Paton Maclay, Sir W. Guy Granet, Gerald A. Steel	December 1921
Committee of the Cabinet Appointed to Examine Part I (Defence Departments) of the Report of Committee on National Expenditure	Winston S. Churchill (chairman), Viscount Birkenhead, E. S. Montagu, Stanley Baldwin, and John Chancellor	January 9, 1922– February 4, 1922; February 23, 1922
Committee on Navy, Army and Air Force Expenditure (Colwyn Committee)	Lord Colwyn (chairman), Lord Chalmers, and Lord Bradbury	August 13, 1925– December 23, 1925
Committee of Imperial Defence: Sub-committee on the Strength and Organisation of the Cavalry	Marquess of Salisbury (chairman), W. S. Churchill, Sir John Gilmour,* Walter Guinness, Viscount Peel, and G. N. Macready Witnesses: Earl Haig, Sir Alexander Godley, Sir David Campbell, Sir Walter Braithwaite, S. C. Peck, and Sir Hugh Trenchard	Proceedings: December 8, 1927– March 28, 1928 Army Estimates: 1927; Final Report: March 1928

*Sir L. Worthington-Evans later replaced Sir John Gilmour

thought and a founding member of the Royal Economic Society, wrote that in the postwar environment it was preferable to be dead than to be seen as a "waster" of the nation's resources.[18] The government formed numerous committees to determine how best to cut national expenditure, including the Committee on National Expenditure; the Committee to Examine Part I (Defence Departments) of the Report of the Geddes Committee on National Expenditure; the Committee on Navy, Army and Air Force Expenditure; and the Committee of Imperial Defence: Sub-committee on the Strength and Organisation of the Cavalry (see table 1).[19] Although all government spending was examined, military budgets were singled out as one of the best ways to save because the cabinet had declared the "Ten-Year Rule," a defense policy stating that the government would not plan to fight a major war within the next ten years.[20] The public and politicians wished to avoid additional continental commitments and to reduce the amount spent on existing responsibilities. Their cost-cutting efforts started with reductions, examining how best to organize their military forces for maximum efficiency with minimal resources. The next step involved examining the ability of new technologies to substitute for older services or to convert existing units into mechanized units.

All military services came under scrutiny for expenditure reductions, but some of the most contentious debates involved the Royal Air Force (RAF) and other aviation establishments. At the end of the Great War, the full potential of airplanes remained under investigation, and agreement on the best organization for air services had not been finalized. The independent RAF proved to be a tempting target for the older services for reduction in an effort to save themselves and perhaps regain their lost aerial assets. How the RAF should be commanded and organized were also universal concerns.

The British had already taken steps to create a more efficient air force during the Great War when they combined the Royal Naval Air Service and the Royal Flying Corps to form the RAF on April 1, 1918. Proponents justified the merger of these organizations by claiming that it prevented competition for resources, particularly airplane procurement, and offered better protection to the homeland, which had been attacked by German airships and heavy bombers during much of the war.[21] The army and navy had demanded more airplanes than producers could supply, pitting the services against one another for airplane procurement. Wartime experience had also demonstrated that divided control of the air (the army was responsible for overland defense, and the navy was tasked with defense over water) greatly handicapped the ability of the rival services' air arms to coordinate responses to hostile aircraft crossing the channel.[22] The creation of the independent RAF in the war's final year quieted some, but it did not end the debate over how to organize aviation most effectively.

Immediately following the war, the War Office and the Admiralty joined forces against the RAF, motivated by a desire to have better control of air assets while, most likely, craving the money allotted to the RAF. Even if they could not claim the RAF's funds, they may have hoped that the savings created by eliminating the RAF would spare them from drastic reductions. They argued that having an independent air service added unnecessary expense because it required the same support services (supply, transport, medical, housing, etc.) as the army and navy. The army and navy maintained that if the older branches reabsorbed the RAF, this duplication could be reduced by one-third, saving the government a significant amount.[23]

The RAF's vulnerability was not lost on its personnel. Rejecting any possible monetary savings by dismembering the RAF, independent air force proponents defended the force's value from an operational and economic point of view. The austere postwar environment, combined with the RAF's desire to maintain its separate existence, led RAF personnel and supporters to sell aviation as a cheaper and more effective way to fulfill the country's responsibilities at home and abroad. Most of the territories in the British Empire were so large that they required individual army units to spread out over hundreds of square miles. Getting troops where they were needed was often an extremely expensive and time-consuming undertaking. While the War Office and Admiralty fought to regain their monopoly on military expenditure, military personnel and politicians outside of the War Office and Admiralty increasingly viewed the airplane, best expressed in the words of historian Keith Jeffery, as "a panacea for the army's problems of imperial security."[24] As early as December 1918, Sir Frederick Sykes, the chief of the air staff, lobbied for the retention of a separate air force, arguing to the secretary for war that "in air power we possess a rapid and economical instrument by which to ensure peace and good government in our outer Empire, and more particularly upon its Asian and African frontiers."[25] Sykes's statement appeared before the economy drive fully developed, but his declaration reflected the justifications for the continued existence of a separate air service used by his successors in the RAF. The argument had two parts: economics and effectiveness. First, he argued that independent aviation was more economical than when aviation was incorporated into other forces. Second, Sykes contended the RAF was an inexpensive and effective service (if not the best) for use in the expanded British commitments. Sykes's 1918 memorandum on South Africa noted, "The moral effect of aircraft on a native population and the great economy as compared with infantry need not be elaborated."[26] Sykes believed the moral effect and savings were so obvious that he did not even bother to make a clear case for either claim.

Sir Hugh Trenchard, Sykes's predecessor and successor, clarified the argument further, along with other supporters of an independent RAF. Trenchard was unwilling to see the newly established branch disappear. During the first few years of his second tenure as RAF chief of staff, he maintained that an independent air force could police the new mandates in the British Empire with air squadrons, a few armored car units, and a few British and locally recruited troops "at a fraction of the cost of a large army garrison."[27] Trenchard proposed that punitive operations normally conducted by ground units of cavalry and infantry would be performed better by RAF planes, which were faster than any surface unit, not restrained by terrain, and had greater firepower than ground-based artillery (large guns are difficult to move away from roads). As a result, each air unit could be smaller than the ground units assigned to the same task, yet they could cover more territory. This concept, "air policing,"[28] seemed like an ideal solution to the expensive problems of empire and also justified the RAF's continued independence and concentration on independent strategic missions.[29] A 1921 article in the *Journal of the Royal United Service Institution* included a more focused attack. A General Staff colonel made a direct assault not just against ground forces in general, as Sykes had done, but specifically against the cavalry. He declared that the terrain in Mesopotamia, East Africa, India, and the Sinai Peninsula "was perfectly ridiculous from the point of view of cavalry reconnaissance" and "in view of the development of aerial reconnaissance, the use of cavalry under such conditions . . . [was] past."[30]

Retention of the RAF through the Promise of Air Policing

By utilizing these arguments, RAF officers received support for the retention of an independent air force from committees formed to examine national expenditure. The belief that the air service could inexpensively police the empire helped the RAF remain an independent service during the British parsimony of the 1920s.[31] As part of its assignment to provide recommendations to the chancellor of the exchequer to reduce national expenditure, the 1921–22 Committee on National Expenditure, perhaps the best-known committee assigned to find ways to reduce spending, examined the merits of retaining or dividing the RAF. This committee—also known as the Geddes Committee after its chair, Sir Eric Campbell-Geddes, a businessman and conservative politician—suggested additional combinations and no separations. Despite the initial public criticism of the members, scope, and possible impact of the committee that accompanied the announcement of its creation, Higgs noted the positive public reception to the report issued by the Geddes Committee.[32] Rather than directly supporting the continuation of the independent air service, the committee concluded that the best solution to increase economy was to place all of the services—army, navy, and air force—under one minister who would coordinate supply, transport, education, medical, and other services.[33] This was not a popular conclusion within the War Office, Admiralty, or the RAF.

The Geddes Committee's decisions did not end debate on whether the RAF should remain independent of the War Office and Admiralty. The RAF's supporters continued to testify before later expenditure committees to suggest other ways to reduce military spending on aviation. Secretary of State for Air F. E. Guest argued before the 1922 committee formed to assess the Geddes Committee's findings that there were two ways to cut aviation expenditure. The first option was to cut Britain's commitments abroad, which would allow a reduction of units. The second choice was to cut army and navy co-operation squadrons, a suggestion already recommended by the Geddes Committee.[34]

Committees throughout the 1920s expressed interest in examining how air policing could save money. The policy promised to be a cheap way to prevent rebellion in places such as Waziristan (on the Raj's northern border at that time, now part of Pakistan) or Afghanistan.[35] In theory, when the inhabitants of a particular area became unruly, the appearance of a few airplanes over their settlements dropping warning leaflets might be sufficient to settle them down. If not, a few well-placed bombs would quiet the area, at least temporarily. Members of the Geddes, Colwyn, and Strength and Organisation of the Cavalry Committees were interested in the possible savings promised by the RAF in reducing the need for large armies at the empire's periphery. The Geddes Committee's December 1921 report noted it was "particularly impressed with the very large saving which we are told can be realised in the Middle East" by utilizing aircraft. Although the report did not identify the committee's sources, it estimated that transferring responsibility from the army to the air force in the Middle East would reduce total military expenditure in the region over 50 percent, from £27 million in 1921–22 to £13 million in 1922–23.[36]

The committee further narrowed its focus from the army to the cavalry, concluding that the RAF could replace the cavalry and produce the nation's desired economy. It stated that the air arm not only added to "the older fighting services" but could substitute for them by "utilizing air forces in place of . . . cavalry in the army" and providing significant economies for the empire's defense.[37] The committee recommended reducing the current twenty-seven cavalry regiments to nineteen with additional possible reductions in the future.[38]

The committee concluded that technologies supplemented manpower, and they could substitute for men. This did not bode well for the older services, especially the cavalry. The retention and possible expansion of RAF units abroad directly involved the reduction of ground forces, particularly the cavalry. Governmental and military attempts to determine how to police the empire continually pitted the cavalry and aviation against one another in the 1920s and 1930s.

The cavalry was not without supporters attempting to save it from massive reductions. In response to the Geddes report, Sir Laming Worthington-Evans, the secretary of state for war, came to the cavalry's defense in the form of a General Staff paper, which circulated among the cabinet. The paper argued that the committee had "advance[d] no valid reason for the drastic reduction proposed in cavalry regiments," although from "the general tenor of their [the committee's] remarks," it was the creation of the air force that brought them to their conclusions.[39]

The War Office admitted that aircraft could discharge some duties more efficiently than the cavalry, such as long-distance reconnaissance, but it argued that to "assert that the place of cavalry can entirely be taken by aircraft in the work of close reconnaissance, protection and support is a complete fallacy." Aircraft were still unproven in traditional cavalry tasks such as advanced and flank guard actions, and the secretary warned that "in the east the need for cavalry will be most urgent, and the committee's proposals" for further reduction of the cavalry would "disastrously cripple the efficiency of our expeditionary force."[40] The War Office agreed with the cavalry reductions that had already taken place.

However, the 1922 Committee Appointed to Examine Part I (Defence Departments) of the Report of the Geddes Committee on National Expenditure criticized the Geddes report, claiming it failed to make the appropriate corrections for inflation, misunderstood the prewar justifications for the army, and did not appreciate the scale of Britain's postwar commitments and responsibilities.[41] Rather than support the Geddes Committee's reductions, the committee suggested that no further reductions should take place and endorsed the proposals made by Secretary of State for War Worthington-Evans, who recommended significant reductions in the army but ones not as drastic as the Geddes Committee's suggestions. For example, the secretary suggested reducing the army by thirty thousand men instead of the fifty thousand advised by the Geddes report.[42]

The interrelated debates over the most economic administration of aviation and substituting air units for ground units to cut expenditure continued in the 1925 Committee on Navy, Army and Air Force Expenditure. Known as the Colwyn Committee after its leader, Frederick Henry Smith, 1st Baron Colwyn and a rubber and cotton manufacturer admitted to the Privy Council in 1924, it supported the 1922 committees' conclusions but more clearly emphasized that eliminating the RAF and returning its airplanes to the army and navy would not avoid unnecessary duplication of support services.[43] The committee affirmed "the necessity for an independent Air Ministry to administer a single, unified Air Service."[44] Secretary of State for Air Sir Samuel Hoare explained the War Office and Admiralty's claim was "inter-departmental warfare . . . waged against the air ministry by the older services." He warned that emulating countries such as the United States, which retained control of the air in the hands of the army and navy, would arrest the RAF's development.[45] This could be particularly dangerous with the amount of British commitments throughout the world. The committee, composed of Lords Colwyn, Chalmers, and Bradbury, expected that substantial savings could be "secured by the extended substitution of air power as a substantive arm." The committee also concluded that far too much was being spent on the cavalry in light of modern technological developments, calling the expenditure "quite unjustifiable." The committee recommended reducing the number of cavalry regiments by "rolling up" existing regiments and reducing cavalry numbers in India.[46]

The discussion of further possible cavalry reductions and conversions into mechanized units continued into the 1927 and 1928 meetings and report of the Committee of Imperial Defence: Sub-committee on the Strength and Organisation of the Cavalry, which was organized to deal with the cavalry of the line. This subcommittee examined witnesses and reviewed the recommendations of previous committees to determine if it was possible to further decrease the size of the cavalry and its services without harming the country's military capabilities. The committee also examined briefly the possibility of converting horse cavalry regiments into mechanized or armored units.

Former Chief of the Imperial General Staff Field-Marshal Sir William Robertson wrote a document in 1927 that was provided to the committee. In said document he argued that no new developments had demonstrated that the cavalry ceased to be essential. He maintained that any additional reduction of the cavalry would be an unjustified risk.[47] A little less than a month before the subcommittee's first meeting, Worthington-Evans wrote a memorandum, later printed for the subcommittee, that stated the army had done more than its fair share to save the empire money. He called the chancellor of the exchequer's suggestion to produce an additional savings of £600,000 by reducing the already barely sufficient twelve cavalry regiments down to six unwise. The secretary recognized that Winston Churchill, the chancellor of the exchequer since 1924 (successively secretary for war 1919–21 and secretary of state for the colonies 1921–22), was under overwhelming and continuous pressure to reduce expenditure, a push that had existed since 1922, but Worthington-Evans argued that the "army has made by far the largest contribution to the reduction in the expenditure on the fighting forces" and that reducing the twelve current cavalry regiments by half would not produce the savings that the chancellor estimated.[48]

As both men were members of the subcommittee, Churchill and Worthington-Evans battled over the degree by which cavalry could be reduced. The first meeting provided the basic path this discussion would take. Churchill adamantly argued that the cavalry be reduced further and attempted to discredit pro-cavalry witnesses even before they appeared in front of the committee. Churchill, the former chairman of the 1922 Committee to Examine Part I of the Geddes Report, took an adversarial position against the cavalry and his fellow committee members. He advocated for the reduction of cavalry even before the committee examined one witness.

In the committee's first meeting, Churchill, not the chairman, took command and stated that he thought the cavalry "furnished the most likely field for economy," and no other reduction in fighting forces could provide any great scope of savings. He also stated self-importantly that "his lead for economy should be accepted" because the government received "all the abuse from newspapers and the economy group in Parliament, as well as from the Opposition." Churchill also recommended that some cavalry regiments should be telescoped (two or more regiments combined into one) and that the number of troops be reduced, because the cavalry was out of proportion to other arms. He suggested replacing cavalry regiments with yeomanry regiments or at least allowing the latter to be established in times of need. He believed yeomanry regiments could "perform the functions of modern Cavalry quite satisfactorily" and more cheaply.[49] In addition, Churchill advocated for transforming cavalry regiments into mechanized units, recommending that the cavalry should be compressed and then converted.

Sir John Gilmour, secretary of state for Scotland and acting secretary of state for war until Worthington-Evans' return from India, cautioned the committee that the British military forces had not "quite reached the stage of mechanical development which enables us to produce a sufficiently trustworthy machine for transformation from cavalry to mechanized units."[50] Another committee member concurred with his colleague, citing statements by senior military officers that Britain was "still very far from the possibility of widespread mechanization."[51] While some reductions were thought to be possible, mechanization was not yet ready to transform the cavalry.

Churchill's animosity was clearly demonstrated when after listening to these concerns, he looked around the room and stated that "the outlook as regards economy looked rather bleak." He claimed the army was not seriously assessing the possibility of creating savings by cutting or transforming the cavalry. He described most generals in high command during the war as cavalrymen who, if questioned, would protest against reduction, insist the cavalry won the Great War, and say that "what we want is cavalry." Churchill had no doubt that the subcommittee could get what he called "cavalry evidence," but he thought it would be harder to "get the other side of the question proved." He maintained that without a reasonable amount of support "from the political side, it was hardly worthwhile pursuing the question."[52] Despite Churchill's skepticism, the subcommittee called witnesses to testify on the possibility as well as the advisability of reducing or transforming the cavalry.

Those witnesses who testified that the horse cavalry was still useful and believed that further reductions would damage it found themselves repeatedly challenged, particularly by Churchill, no matter their military credentials. Churchill acted as if any former contact with the cavalry made officers incapable of making a fair assessment of the branch's future. Even before the Sub-committee on the Strength and Organisation of the Cavalry examined the majority of witnesses, he questioned the partiality of the witnesses because, he claimed, military leaders were "mostly cavalrymen."[53] Once the witnesses arrived, Churchill aggressively asked simplified and leading questions regarding which military branch should be reduced, trying to force them to advocate for reducing the cavalry. When witnesses refused to agree to even theoretical reductions in the cavalry before all else, Churchill made more ridiculous and oversimplified comparisons to get them to put the cavalry on the chopping block. Because they did not take his bait, Churchill maintained his original position that the witnesses would always defend the cavalry. During testimony at the subcommittee's third meeting, Churchill finally heard what he wanted to hear from a military officer: regular cavalry could be converted into a mechanized form without forming new units. After Churchill heard this statement, he believed that little more committee research was necessary.[54]

When Worthington-Evans returned from India to resume his position as secretary of state for war and to attend the subcommittee's fourth meeting, Churchill's position was unyielding. Although this was the first subcommittee meeting he attended, Worthington-Evans was already familiar with and objected to Churchill's tactics in earlier meetings. During discussions on what to include in the subcommittee's final report, Worthington-Evans accused Churchill of asking "trap" questions to make the witnesses appear absurd or put them in a corner. Worthington-Evans objected to Churchill's suggested inclusion of witnesses' specific answers to these questions in light of the nature of the questions posed. He thought they would not be accurate statements of what the witnesses actually believed. Other members provided a moderating voice, suggesting using more general comments about witnesses' testimony. Churchill strongly objected but was overruled. He refused to sign the committee's final report. Instead, he added a note to the end of the report, listing his objections and recommendations.[55] In the end, Churchill got what he desired from the sessions—support for additional cavalry reductions.

British Air Policing in Practice

Although the national expenditure committees accepted the RAF supporters' argument that air policing, using airplanes to replace ground troops, would be cheap and effective, this did not work out in practice. The British Air Ministry argued that campaigns throughout the empire demonstrated the effectiveness of air policing.[56] Operations in Somalia, Mesopotamia, Palestine, Aden, India, and other areas throughout the 1920s and 1930s were used to strengthen their claims.[57] Yet historians of air policing agree that those who argued that airpower was successful on its own were wrong.[58] Despite the impression left by RAF reports, "most of the operations in the colonies in the interwar years were in support of, and in cooperation with, ground troops."[59] When working with other arms, aircraft conducted reconnaissance, artillery observation, bombardment, convoy protection, and casualty evacuation operations. In addition to those tasks, aircraft led machine gun raids, supply operations, rebellion-deterring demonstrations, and communications missions.[60]

RAF reports initially acknowledged this cooperation between ground and air units, but its public statements about the vital contributions of surface forces "gradually faded" in its accounts of operations in the empire, according to historians James Corum and Wray Johnson.[61] The increased absence, accidental or not, of this information created a belief in Britain that airpower was a successful precision instrument on its own. Yet even the air staff never claimed that aircraft would completely replace ground forces in colonial and mandated territories. They did not deny that ground troops were required to defend air bases and political centers.[62] This was reflected in the Sub-committee on the Strength and Organisation of the Cavalry, which stated, "It is not suggested that by reason of this replacement [RAF for cavalry], the establishment of ground mounted troops of one sort or the other could be reduced."[63]

Numerous operational limitations hurt the ability of air policing to succeed without ground support. In Iraq, airplanes experienced serious problems. Engines overheated, "propellers warped, tyres were punctured by thorns and shock absorbers perished."[64] Inclement weather over the mountains along the Afghan frontier hindered operations in the Third Afghan War in 1919.[65] On the northwest border of India, a "month-long bombing campaign showed that the tribesmen could adapt to aerial attack."[66]

British cavalrymen questioned the empire's seeming overreliance on aircraft and praised combined operations.[67] An officer of the 12th Royal Lancers charged that operations in Morocco, Syria, and the Rif demonstrated that "neither mechanical vehicles nor aircraft, whether acting alone or in combination . . . [could] deal satisfactorily with a cunning enemy in difficult terrain."[68] He also rejected the ability of these forces to act as adequate police forces throughout the empire, arguing that airplanes by themselves could not have prevented violence in the Hankow and Shanghai concessions as ground forces had.[69]

The cooperative nature of successful operations was apparent in the first use of airpower on the British frontier in 1917, when ground troops and BE2c biplanes worked together during the Waziristan campaign in northwest India. The effects of airpower were transitory, but they were more lasting and successful when used in conjunction with ground forces. Within two years, ground and aerial forces were used again with improved airplanes, contributing to the amir's decision to sue for peace.[70] In Egypt in 1919, the RAF also worked with ground troops while they "patrolled communications, scattered propaganda, delivered mail, relieved garrisons," and sometimes attacked Bedouins and demonstrators.[71] Historian Philip Anthony Towle argued that a "combination of air and camel power" destroyed opposing forces in Somalia.[72] The 1920 rebellion in Iraq also persuaded observers that aircraft were unable to take the place of ground troops as the main imperial police force.[73] However, economics overruled effectiveness, and the RAF's offer to garrison the country at a minimal cost was quickly accepted by the British government because of the great expense of Iraq in personnel and resources.[74]

Economic constraints also hampered effective combined operations. The use of aviation in the empire proved that airplanes were, according to Corum and Johnson, "tremendously effective as a force enhancer in military operations." Limited funding, however, complicated the RAF's ability to function in both its support and independent roles.[75] When budget reductions were demanded, the RAF did not want to diminish its ability to conduct independent operations. Instead, RAF officers argued that cooperating with ground forces and providing cooperation squadrons were not the RAF's first or most important priorities. Therefore, if further reductions were required, army and navy co-operation squadrons would be eliminated—and they were. The RAF drastically reduced its army and navy co-operation squadrons, but air personnel were unwilling to reduce the number of independent (strategic bombardment) squadrons that justified its separate existence. If the RAF had to lose funding, it preferred to eliminate units assigned to cooperate with ground forces.[76]

This reduction supported aviation's visions of independent future missions at the expense of ground unit's visions of cooperation. The "army emphasized a continental commitment and 'small wars,' while the air force stressed home defence and the independent bomber offensive." When the army requested thirty more cooperation units from 1923 to 1924, the RAF provided only four.[77] Historian Richard Muller interpreted this to mean that the RAF was not concerned about preparing for possible future wars; instead, it was only working to insure its survival as an independent branch in the austere environment of the interwar period.[78]

This conviction contributed to the end of discussions about the return of RAF assets to the two elder services and led to the emergence of a myth about the success of air control in the empire. Historian David Omissi explained that this myth maintained that "a small force of airplanes was cheaper and more efficient" than ground forces because it could produce equivalent amounts of disruption and destruction, the goal of punitive missions. He argued that accurate explanations of why operations succeeded "became irrelevant." The cabinet continued to be "swayed by the impressive economies promised by Churchill" and convinced themselves that the myth was true.[79]

In addition to its lack of long-term effectiveness, substituting aircraft for ground forces did not produce the promised economy. Secretary of State for the Colonies Leo Amery wrote to officers in Palestine in 1926 that instead of producing savings, two RAF flights in Palestine would cost £110,000 more than the cavalry regiment they were scheduled to replace. He recommended that Palestine should contribute additional resources to produce savings instead.[80] The cabinet decided to remove the cavalry regiment anyway because it would be cheaper to maintain it at home, not because aviation would be cheaper.[81]

American Military Expenditure Debates

Similar to their British counterparts, Americans wished to decrease military expenditure while maintaining military effectiveness. American officers also wished to incorporate new technologies, such as aircraft and tanks, more fully into their forces. However, the United States, unlike Britain, did not need to police a large empire. As a result, it seemed unnecessary to squander the United States' resources on a large military establishment, especially not on obsolete branches and unnecessary equipment.[82]

The American public made their position on waste clear to their representatives after the postwar "billion-dollar bonfire," when over one hundred surplus obsolete aircraft, some never used, were burned in Europe to spare the government the expense of shipping them back to the United States. The American public was horrified by the destruction of so many expensive aircraft. Politicians were equally shocked.[83] Although initial reports proved to be misleading about the type, amount, and reason for the equipment's destruction, Senator Henry Cabot Lodge of Massachusetts, chairman of the Senate Foreign Relations Committee and opponent of Woodrow Wilson's League of Nations, used the bonfire as evidence of the "extravagance and waste" of the Wilson administration. Interestingly, Lodge mentioned to an audience the number of horses the army had acquired during the war as further evidence that President Wilson was wasting the public's money, which elicited laughter from the assembled crowd.[84] He compared the money wasted on supporting animals and the money lost when planes were destroyed. Horses were increasingly viewed by the general public as unnecessary. Americans did not want to spend money on the military or on horses, which was not a good sign for the continuation of the horse cavalry.

Although both countries had the similar goal to produce economy and effectiveness of military forces, British and American debates diverged significantly. While the composition of air forces also proved a contentious issue in the United States, the primary target of the independent air force advocates was not the cavalry but the navy, arguing that they could complete the missions of an older service better and cheaper. This divergence can be attributed to the unique military challenges the two nations faced.

Unlike Britain, which had strong potential enemies equipped with air forces across the English Channel and North Sea (Germany and France) and significant overseas commitments, the United States had no serious aerial challengers along its land borders and limited responsibilities beyond its own borders. The possible threats to the United States were more likely to come from the sea and its southern border, at least until airplane technology advanced further. Mobile horse cavalry remained vital for patrolling the long Mexican border. Although the value of new technologies spurred the postwar creation of the Air Service and Chemical Warfare Service, postwar economy cut the army to less than 130,000 men who had to survive off Great War surplus supplies and equipment.[85]

Unsatisfied with the size and composition of the air force, aviation's supporters in the United States promoted the development of a new aviation branch as a way to maintain and improve military effectiveness and efficiency. General William "Billy" Mitchell, who had commanded all of the American aerial combat units in France during the Great War, was the most well-known, outspoken, and controversial advocate of a separate and independent air service. Attempting to spur changes in American air force and funding structures, Mitchell attacked the Navy and War Departments in the press for poorly handling the national defense.[86] In his many publications and speeches, including *Our Air Force: The Keystone of National Defense* and *Winged Defense: The Development and Possibilities of Modern Airpower—Economic and Military,* Mitchell stressed the importance of an independent organization. He argued that an independent air force would need to provide only moderate support for ground and naval troops because it was capable of winning a war almost by itself.[87] In that case, it did not make sense to keep American air services divided between the army and navy.

While their justifications were not primarily financial, Mitchell and his supporters attempted to take advantage of the parsimonious atmosphere, arguing that airplanes would be a cheap replacement for older military technologies and services. He argued that "the much more effective and economical airplane had dethroned the battleship as the queen of national power."[88] He utilized the 1921 service tests evaluating the damage aircraft could inflict on battleships as evidence of the strength of an independent air force and the weakness of ships at sea. Numerous vessels were sunk over two months in the summer of 1921, but the July 18 experiment utilizing the *Ostfriesland,* a German battleship taken in reparations, garnered the most attention. After two days of attack, the *Ostfriesland* sank into the sea. General Mitchell cited the test as proof that there was no reason to devote funds to build new battleships made obsolete by the airplanes' ability to sink them easily.[89] Instead, the money could be saved or used to further aviation development.

The U.S. Navy responded directly to Mitchell's assertions, accusing Mitchell of invalidating the test by violating the rules, and then drawing incorrect conclusions from the evidence.[90] Sinking an unmanned stationary naval vessel with bombs dropped from unopposed airplanes did not adequately simulate an actual naval battle. One congressman stated that the test was equivalent to shooting a "tiger in a cage at the zoo" because it did not really demonstrate what it would be like to shoot that same "tiger in the jungle."[91] Instead, the navy concluded that the test proved the "utter and absolute interdependence of aircraft and surface craft."[92]

Mitchell, undeterred by this setback, continued agitating for an independent air service, repeating his faith in the airplane as an instrument of war while adding a financial element to his rhetoric. The economics of Mitchell's proposals are best expressed in his memoirs, published in 1928. He argued that airpower would allow "a few men and comparatively few dollars" to bring "about the most terrific effect ever known against opposing vital centers."[93] Bombs and fire from the air combined with chemical weapons would "unquestionably decide a future war." Mitchell wrote, "Today, armies and navies are entirely incapable of insuring a nation's defense."[94]

When it came to aviation's impact on the nation's security policy, defined by historian Brian McAllister Linn as "noninterventionist, deterrent, and focused on continental defense," aviation advocates' primary army target was the coast artillery. This branch, responsible for manning coastal defenses and protecting the nation's borders, at first welcomed aviation's assistance but later saw it as a challenger to its existence.[95] Although the navy and coast artillery took the brunt of Mitchell's attack, the related arguments about reducing military expenditure also affected the army and cavalry. His opinions infected the public, politicians, and some military personnel, but the United States did not grant the U.S. Army Air Force its independence until it proved its value during World War II. Hoping to prevent arguments that a buildup of aviation should come at the expense of the cavalry, Major General Fox Conner, the acting chief of staff of the army in 1926, stated that the air corps should not come at the "expense of any other branch" and advocated that pending Army Air Service budget bill, H.R. 10827, should "provide that that shall not be done." He stated policing the two-thousand-mile-long southern border of the United States still required the cavalry. Connor rejected the suggestion of following the example of European nations and creating an independent air force because he believed the situation in the United States was different from that in Europe, particularly in Britain and France, both of which had powerful air forces dangerously close (Italy and Germany).[96]

Incidents on the border had already demonstrated the value of mounted troops and the limits of aviation. The Americans had first used their airplanes on their own frontier during the 1916 Mexican Punitive Expedition. In 1916, the 1st Aero Squadron flew 540 sorties and more than 340 flying hours over 19,000 miles, supporting ground troops from March to August with eight Curtiss JN-3s. Popularly known as "Jenny" due to its designation, the JN-3 was a two-man biplane with a maximum speed of less than eighty miles per hour and a top altitude of only ten thousand feet. Although the JN-3 was an adequate trainer, it proved "woefully inadequate as a combat platform."[97]

The planes experienced numerous mechanical problems throughout their brief service in Mexico. The harsh climate required removing their laminated wood propellers after each flight for storage in humidors to prevent cracking. The terrain also caused problems as the JN-3s were incapable of going over even many of the foothills of the Sierra Madre. The passes and gorges also contained "high winds, vicious cross currents, and downdrafts" that made piloting the airplanes difficult. The squadron was eventually relegated to carrying mail and maintaining communications with troops in forward positions. The JN-3s' replacements, Curtiss R-2s, which were little more advanced than the Jenny, hardly improved the squadron's capacities as they suffered "from engine problems and structural defects."[98] All these problems in 1916 made aircraft incapable of policing the border unaided by mobile horse cavalry.

After the Great War ended, the army continued to monitor its southern border with both air and ground components. In the Big Bend of Texas, the 5th United States Cavalry Regiment patrolled on the ground while two squadrons assigned to it flew circuits two or three times a day, watching the border for any disturbance. This Army Air Service border patrol had been formed in late 1919. Aviators did not call for an end to cavalry operations along the frontier. Both aviators and cavalrymen recognized the value of air and horse units working cooperatively to accomplish border security. The office of the chief of the cavalry recognized the great value of aviation to patrol the border alongside a large cavalry force.[99]

Border service was not the cavalry's only responsibility, and not all cavalry officers were happy about the cavalry being stationed primarily on the border. Numerous letters were sent among cavalry commanders in the 1930s commenting on the number and placement of cavalry units. In 1930, Colonel J. R. Lindsey, previously serving with the 15th, 11th, and 14th Cavalries, recipient of the Distinguished Service Medal during the Great War while serving with the 328th Infantry Regiment, and at the time chief of staff of the 61st Cavalry Division, Organized Reserve Corps (1928–32), wrote to Major George S. Patton that keeping the cavalry in the south in case of problems with Mexico was a mistake.[100] He stressed that horsed cavalry was needed in other locations for vital training. He expanded further, attacking the War Department's established policies as "punk and should be knocked into a cocked hat." Lindsey asked whether the army was "going to let a petit pais like Mexico dominate the permanent distribution of our regular army?"[101] Patton forwarded the letter to Chief of Cavalry Guy Henry. Lindsey's correspondence produced a series of letters that ended up going beyond his initial displeasure with cavalry unit locations.

In his response to Lindsey, Henry expressed his own frustration about cavalry numbers and locations, seemingly suggesting that Lindsey was not addressing the correct person. Henry discussed the complicated situation between the increase in aviation and the reduction of the cavalry, exactly what Fox Conner had hoped to prevent four years earlier. Henry commented that upcoming new air corps increments, which were the periodic disbursement of appropriated funds, would be "as everyone knows, made at the expense of other branches of the service." Henry stated that the outlook for the cavalry in 1930 was getting worse.[102] In November 1930, the cavalry "suffered a reduction of 343 men for the 4th Air Corps increment," and War Department plans included an additional cavalry cut for the development of the fifth air corps increment in the near future.[103] Henry tried to stop cavalry reductions by encouraging corps area commanders to request additional troops for summer training. Henry sent letters or radiograms to the commanders of the 3rd U.S. Cavalry Regiment in Baltimore, Maryland, the VI Corps area, and the IV Corps area as well as additional letters to the adjutant general that stressed the importance of these commanders' requests.[104] Henry confided to Major General Frank Parker that he campaigned to elicit requests for additional cavalry from commanders because "special effort was being made to get back the men which have been lost by the various arms to the air corps, and with your request in, the cavalry would undoubtedly have obtained its use percentage of these men." Yet despite his attempts, Henry lamented that his efforts were in vain, adding "unfortunately, this, I understand, was knocked out yesterday in the budget."[105]

Some cavalrymen continued to attempt to get their numbers increased by arguing they were needed outside the United States. They argued that the cavalry should be part of the American expeditionary forces outside of the continental United States because they were more capable of completing certain necessary tasks than air and infantry units. In a letter to the adjutant general on July 31, 1934, cavalry Colonel G. Williams recommended assigning horse cavalry to the Panama Canal Department because they were much better suited to accurate observation and identification than aviation. Williams echoed arguments made in previous decades: "Mounted men only could make positive reconnaissance and obtain definite information." Aircraft could not locate "bodies of enemy troops in the close country of the Isthmus or on the hundreds of trails which exist in the jungles."[106] Unfortunately, Williams lacked strong support from the commander of the Panama Canal who acknowledged that although a cavalry squadron would be a "convenient addition," other reinforcements had a higher priority, particularly infantry.[107]

Despite the lack of full support from the area commander, Williams wrote Chief of Cavalry Major General Leon B. Kromer to argue that the Canal Zone was the best place to demonstrate the cavalry's continued value and that it was the duty of all cavalry officers to "keep the cavalry, horse, in the foreground showing its necessity."[108] Kromer agreed that cavalrymen should keep horse cavalry in the foreground in both word and deed, but he noted that any change to the status quo would need to overcome powerful economic considerations. He also asserted that an increase in cavalry could only occur if a study of the whole defense organization showed such action necessary.[109] Unfortunately for the horse cavalry, no such conclusion was made.

⁓

Concerns about the postwar economy and beliefs in the abilities of new technologies utilized in the war to replace the old were too much for the horse cavalry to overcome, and its size and significance shrank in the interwar period. Despite the large initial investment required to incorporate new technologies into the military establishment, the promised long-term savings in manpower and support services for men and animals tempted politicians to increase the use of airplanes at the expense of older services. The fact that the air forces could not live up to expectations, managing only mixed success in imperial and border work, did not seem to matter because the myth of airpower won.

In Britain, the debates about the purpose, organization, and ability to substitute planes for the older services lasted throughout the 1920s and 1930s with the status quo prevailing. The RAF remained autonomous and concentrated on further developing independent actions while reducing the number of Army co-operation squadrons to appease the government's desire for economy. As a result, the cavalry had fewer opportunities to work directly with aviation to develop tactics and doctrine that would strengthen the cavalry or even be used to fulfill the cavalry's duties.

Although cavalrymen argued for their branch's ability to transform, their efforts to demonstrate the cavalry's progressive character failed to prevent reductions to their branch. British cavalry regiments were reduced by combining them, and others would be converted to armor units (as will be discussed in more detail in chapter 5). There had already been a reduction from a high of twenty-seven regiments down to nineteen in 1921 and additional reductions to twelve in 1927. American horse cavalry regiments remained in the 1920s because the defense of the southern border still required them. Although the United States retained horse cavalry, their effective number shrank. Reductions of more than 340 men were directly attributable to aviation.

Of all of the arguments against the cavalry, economics proved the most successful. Despite support from military officers and members of the government for their continued value, horse cavalry numbers shrank as tanks, armored cars, and aviation slowly altered the composition of cavalry units. Cavalrymen and cavalry supporters maintained that these other technologies were not yet adequate in the 1920s and early 1930s. Horse cavalry was retained but in a reduced role. Cavalrymen did not universally or completely oppose the changes; instead, they actively participated in the orchestration of these changes and continued their attempt to deepen their relationship with aviation and other technologies as the next chapter will show.

CHAPTER FIVE

Autogiros and Mechanization
The 1930s

By the late 1920s, aviation had appropriated many of the cavalry's traditional reconnaissance roles, but it had failed to replace the cavalry entirely. Aviation had altered the roles and employment of the cavalry, a fact made apparent in peacetime exercises and officially endorsed in both British and American policy. Cavalrymen accepted aviation's assumption of the cavalry's more burdensome duties and worked on creating the best possible cooperative relationship with air forces throughout the 1930s. Even as the cavalry tried to fully integrate aviation and create organic aerial capabilities, the 1930s proved to be a decade with new aviation challenges and ground threats that targeted the branch's foundation: its identity as horse-mounted soldiers. This decade contained the extremes of cavalrymen working to develop their own aircraft while defending the continued vitality of the horse over machines.

This chapter starts by describing the stabilized cooperative relationship between aviation and cavalry as seen in published regulations, textbooks, and school curricula throughout the 1930s. It then examines problems that occurred in implementing these policies in training exercises when aviation priorities shifted away from working with ground units. The cavalry contin-ued to express a desire to work with aviation, a desire increasingly refused by more and more politically powerful aviation supporters. When their requests failed to produce the desired results, cavalrymen attempted to gain control of their own aircraft units and aircraft. The United States and, to a far lesser extent, Britain, saw cavalry trying, albeit unsuccessfully, to establish organic air arms and promote the new technology of the autogiro. However, finances and the outbreak of World War II brought the experimentation to an end.

At the same time that cavalrymen tested the autogiro, mechanized ground vehicles challenged the cavalry's traditional roles and its primary method of transportation, the horse. Cavalrymen responded to the chal-lenges of mechanization in the form of tanks and armored cars in much the same way they had initially reacted to aircraft, though in a much more compressed timeframe. This chapter surveys the similarities between the responses utilized by cavalrymen in response to aviation and to the newer challengers. Despite the similar tactics, cavalrymen could not minimize the impact of mechanization on their branch, and horse cavalry began disappearing at an accelerated pace. Mechanization provided the opportunity for the cavalry to retain its roles and élan but only by transforming itself from a horse-based service to a machine-based one.

Doctrine and Regulations

By the early 1930s, the relationship between aviation and the cavalry had stabilized. Official doctrine had established and defined the separate and cooperative duties and roles of air- and horse-mounted units. Military publications, military school courses, and maneuver reports disseminated these ideas. These roles were similar in the United States and Britain despite their different national commitments and challenges.

Field Service Regulations of the United States 1923 (FSR-US 1923), described by William Odom, author of *After the Trenches,* as "the single best available description of how the army believed it should wage war," had already established the cooperative relationship between the air service and the cavalry.[1] *FSR-US 1923* stated that ground reconnaissance supplemented aerial observation as well as utilized intelligence provided by aviation units to determine dispositions for its own reconnaissance. If aviation proved unable to make this initial reconnaissance, however, the cavalry had to be "prepared to extend its reconnaissance to secure all information desired" and pass that information along to all interested parties.[2] *FSR-US 1923* identified the cavalry as the "principal agency of terrestrial distant and close reconnaissance," but noted that the Army Air Service also participated "in all phases of distant, close, and battle reconnaissance."[3] These regulations were not superseded until 1939, but the new regulations retained much of the previous doctrine. The 1937 *Tactics and Technique of Cavalry* also clearly outlined this cooperation, noting "aviation gains information of a general nature; cavalry obtains specific information to supplement that of the air force."[4]

As in the 1920s, the benefits of cooperation between aviation and the cavalry appeared in student papers, textbooks, and lectures at army graduate schools. In his 1936 Command and General Staff School student paper, a Quartermaster Corps captain described the ideal relationship between aviation and the cavalry as one in which each arm enhances and aids the other, concluding that a close cooperative relationship between them was crucial.[5] The 1937 text *Reconnaissance Security Marches Halts*, utilized at the Command and General Staff School, taught students, including cavalrymen, that commanders, or members of their staff, would frequently utilize airplanes to conduct reconnaissance to gather necessary information for successful operations, such as the nature of the terrain. However, the characteristics of each situation would determine whether airplanes or other means of transportation were used.[6] The Fort Riley Cavalry School taught students about aviation's capabilities and uses and trained them to appreciate the value of aerial assistance.[7] A major and instructor at the Cavalry School argued in the *Cavalry Journal* that it would not require much effort "to imagine large cavalry units of the future preceded by aviation on long-distance reconnaissance" because the information gained by aviation should simplify the "engagement of the main columns of the cavalry command."[8]

The air service could provide the most distant, or strategic, reconnaissance and was thus "the principal agency for seeking information within the enemy lines, for quickly verifying reports of enemy activities, or for meeting emergency needs for reconnaissance at a distance." The cavalry, in contrast, could not "be expected to gain information within the area occupied by its attached air service."[9] The cavalry's task was to establish and maintain contact with hostile forces, something aerial units could not do.[10] The 1939 *Reconnaissance by Horse Cavalry Regiments and Smaller Units* noted that the mounted reconnaissance patrol was to be utilized to augment information obtained by other ground units and the Air Corps.[11]

This collaborative relationship was also reflected in the curriculum and lectures of the Air Corps Tactical School at Maxwell Field, Alabama, during the mid- to late 1930s. The school's main mission was to train aviators in the "strategy, tactics, and techniques of airpower."[12] The school also taught infantry, cavalry, and combined arms courses to educate airmen in the organization, characteristics, and roles of army branches as well as how to cooperate with these units. The subjects explored in its cavalry courses included offensive combat, reconnaissance, pursuit, cavalry organization, cavalry characteristics, cavalry roles, and delaying action.[13] Cavalry officers served as instructors for these courses, and the texts reflected a combined arms doctrine.[14]

Lectures at the Air Corps Tactical School reflected current doctrine and tactics. The courses were intended to educate students about how cavalry could or could not and should or should not be employed. Cavalry courses taught that there were duties best performed by aviation and some that could only be accomplished by the cavalry, either horse or mechanized. Aviation could provide early information regarding the enemy's disposition and movements, and the cavalry could utilize this data to determine its dispositions for more detailed reconnaissance. The cavalry would provide all reconnaissance in situations where hostile aviation or weather prevented aerial craft from completing this task. As one instructor stated, "Each branch supplements and assists the other, the most intimate team play between them is essential." Each reconnaissance agency "should be employed in accordance with its characteristics."[15] The same instructor argued in a later course that while aviation had partly taken over the cavalry's duties, "there [was] plenty left for the most mobile of our ground forces (cavalry)."[16]

The most important characteristics of the cavalry were mobility, firepower, and shock. Horse cavalry was useful in terrain unsuited to motor vehicles, and shock was still a possible application of squadrons or smaller units. However, Lieutenant Colonel J. C. Mullenix, the instructor of the 1938–39 cavalry course, argued that weapons should be used only when they were well-suited for the situation. Just because bombardment aviation was available did not mean that was a "good reason to order it to bomb," nor was it appropriate to use the cavalry to breach fortifications or gallop arbitrarily simply because it was present. Although they were still mentioned, shock tactics were not central to the course curriculum, and they were excluded entirely from the curriculum in some school years.[17]

In describing the cooperative relationship between the cavalry and aviation in the American *Cavalry Journal*, cavalrymen continued to point out that the loss of some of their traditional roles to aviation only strengthened the cavalry. Brigadier General Hamilton Hawkins, one of the foremost authorities on cavalry tactics, maintained that "now that the distant reconnaissance is largely taken over by the air force, the cavalry is relieved of a great part, though not all, of this exhausting duty and can be better saved for its great roles."[18] Even in 1938, the *Cavalry Journal* included a statement from the late Commander in Chief of the German Army General Hans von Seeckt that "the AVIATOR has come to the aid, not to replace the cavalry" and that "close reconnaissance is left to the cavalry whose vision is not dimmed by clouded skies."[19] These assessments did not invalidate the aircraft's value, but they reinforced the idea that aviation was part of a mixed force that included the cavalry. Lieutenant Colonel Mullenix noted with gratitude that aviation had "relieved cavalry of certain of the grueling, horse-destroying, man-killing distant reconnaissance that formerly frequently put large cavalry units . . . practically out of business before the fighting started."[20]

Policy did not vary greatly across the Atlantic. British *Field Service Regulations 1930* claimed that their doctrine was unique because the problems incurred in an empire of "self-governing communities widely separated and of varying resources" were "peculiarly its own." The British military had to be ready for anything "from a small expedition against an uncivilized enemy to a world-wide war." Therefore, its planning had to "ensure elasticity, unity of effort, decentralization of control, and economy."[21] Their self-proclaimed uniqueness, however, did not create a tactical doctrine distinct from the Americans for the cavalry and aviation. British regulations also preached cooperation between the cavalry and air services. The 1931 British *Cavalry Training* stated clearly that "an army can exert its full power only when all its parts act in close co-operation."[22] This point was deemed so important that it appeared in bold print:

> Throughout their training, therefore, all ranks of cavalry must be taught to realize the close relationship between their own role and that of the other arms in battle. They must understand the methods employed by infantry, artillery, engineers, tanks and aircraft to support them, they must appreciate the importance of close liaison and intimate mutual co-operation during the preliminary arrangements for a battle and throughout every stage of the action.[23]

British regulations also outlined the individual duties of the cavalry and aviation. The 1935 *Field Service Regulations* identified cavalry duties as reconnaissance and force protection with the additional battle duties of delaying the enemy, safeguarding the flank, forming a mobile reserve, carrying out pursuits, covering withdrawals, and conducting special missions.[24]

The 1935 regulations also divided information and reconnaissance duties between air and ground units on the basis of their differing capabilities. Air reconnaissance could quickly gain "information of the enemy's movements and of the topography of the country . . . to a great depth"; however, it remained limited by weather and terrain and could not "produce the detailed information necessary for tactical plans."[25] In these situations, ground units had to supplement aviation, particularly in the forward battle area. The 1935

Field Service Regulations clarified these duties further by providing operational specifics for higher formations: "strategical reconnaissance [would] be mainly the work of the air force" but could be "supplemented and confirmed by mechanized forces or mounted troops according to the suitability of the ground for their operations."[26] Further regulations named the type of aerial elements that would work in cooperation with ground forces. During battle, Army cooperation squadrons would "carry out the duties of medium, close, artillery and photographic reconnaissance." In addition to gaining air superiority, aircraft in the British Army co-operation squadrons were "specially trained for work with the army," with their principal tasks being "reconnaissance, including photography, and artillery observation."[27]

Making the Relationship Work

The formal doctrine, written regulations, and school curricula did not end cavalry debates over the use of aviation. Making a cooperative relationship work demanded serious effort, especially as aircraft continued to progress technologically and aviation proponents shifted their attention toward independent activities. Training and maneuvers in the 1930s continued the work of the 1920s, testing and practicing joint operations with airplanes and the cavalry. The cavalry in both the United States and Britain had become dependent on the assistance of airplanes. The cavalry did not just want an aerial branch in front of it but an aerial branch working closely with it in its missions, as described in doctrine.

Both British and American training emphasized the importance of cooperation. In maneuvers, aviation supported the cavalry in reconnaissance, attack, and communications. Reports on the May 1930 U.S. Cavalry-Infantry field maneuvers near San Antonio, Texas, included praise from the cavalry for aviation's ability to assist their branch.[28] The 1931 maneuvers of the 1st Cavalry Division also stressed teaching "sound military lessons" about the joint employment of forces and not the "superiority of one force over another."[29] In Great Britain, the 1930 annual individual and collective training seasons also tested how air units cooperated with ground troops, particularly in tactical reconnaissance.[30] During the 1932–33 individual training period, several exercises in Great Britain and Egypt tested cooperation between ground forces and Royal Air Force (RAF) bombers, fighters, and Army co-operation squadron aircraft. The chief of the Imperial General Staff concluded in the memorandum resulting from the collective training period that the principles for "the employment of air forces by the army" were "sound."[31] The report on the 1934 British training season noted, "Air reconnaissance is important in every phase of the operation, and the co-operation of bombers and fighters in assisting in the protection of the mobile force, either directly or indirectly, may be a factor vital to success."[32]

Reports from maneuvers, however, were not entirely positive. Some observers worried that the artificial conditions set by organizers created false lessons. Major Patton argued that the information provided by aircraft during the 1929 cavalry division maneuvers was not incomplete or lacking but was "too good." The unrealistic test conditions produced an "undue reliance" on messages dropped from aircraft, which could only be corrected by restricting the employment of aircraft on maneuvers.[33] A cavalry major and instructor at the Cavalry School also noted that the "machine auxiliaries," airplanes and armored cars, still had "definite limitations, based upon weather, terrain, mechanical difficulties, and supply."[34] Cavalry observers of maneuvers in 1935 also questioned the value planes provided, noting that aviators put their planes at risk by flying too low, which, they argued, would make them easy targets from ground fire.[35]

Although some cavalrymen still mentioned aircraft's limitations, cavalry dissatisfaction about aviation during maneuvers in the United States stemmed more from misuse or absence of desired air units than with their limitations. In the early 1930s, the cavalry lacked enough planes at its Fort Riley maneuvers to conduct proper cooperation. In the words of air historian I. B. Holley, there was "minuscule aerial participation" in the 1932 maneuvers.[36] Three years later, 6th Cavalry Lieutenant Colonel Kinzie Edmunds and Captain Rufus Ramey still complained that a shortage of planes made training with the Air Corps in the May 1935 maneuvers "not satisfactory." They had only a few observation planes with which to experiment. As a result, the cavalry lacked sufficient aerial reconnaissance abilities and did not have a chance to test themselves against the attack aviation of their mock opponents.[37]

The lack of sufficient numbers of airplanes for cooperation in maneuvers was evidence of a far more troubling situation. In both the United Kingdom and the United States during the 1930s, the army's relationship with aviation changed drastically when advances in aviation technology made large-scale strategic bombing, a concept introduced during the previous decade, practicable. Both the RAF and the United States Air Corps eagerly shifted their focus from the tactical support of ground troops to independent strategic bombardment, changing the tone and content of the cavalry debates in the 1930s. As discussed in chapter 4, following the Great War the RAF decided to "concentrate on the development of a force which could operate primarily at the strategic end of the air power spectrum."[38] There were two reasons for this decision. The first was to justify the maintenance of the independent RAF, and the second was to minimize military expense in the frugal post-war environment. The second was accomplished in part by eliminating or reducing army and navy co-operation squadrons.

In the United States, Air Corps leaders conducted a similar discussion on the role of aviation during the 1920s, but they did not follow the British example until the 1930s. The basic ideas behind the creation of the air force, according to an Army Air Force historian, were founded on the three basic ideas on the role of air power in national defense: "(1) air power to be effective must be based on bombardment; (2) command principles should be established by which that bombardment could be directed against proper targets; (3) heavy bombers of sufficient range should be constructed so that the doctrines might be implemented under the peculiar geographical conditions affecting the United States."[39] In a decade of limited budgets, both the British and American air arms determined that they could not focus on too many roles, and as strategic bombing promised to liberate them from too close a relationship with land operations, their leaders focused on preparations for independent operations.

Aviation historian Roger Connor observed that "with the rise of airpower advocates" who desired an independent aviation service, "the Air Corps saw that only by increasing the capability of its technology could it achieve military objectives on par with land and naval forces." As a result, the U.S. Air Corps' Material Division "saw a mandate to create higher, faster aircraft with ever greater payloads and range."[40] The technological seductiveness of the faster and bigger, shinier and newer did not help the army in its efforts to retain or gain tactical air support units. As the speed and size of airplanes continued to increase and strategic air power doctrine became more institutionalized, ground forces lost military support from air components and moral support from air leaders. Instead of defending its continued value in reconnaissance against aviation, as it once had to, the cavalry repeatedly had to make special efforts to request its affiliated air arms to continue cooperating.

Once they adopted this new doctrine, RAF and U.S. Air Corps leaders increasingly neglected tactical operations in support of ground units as well as the development of the types of aircraft needed to complete those missions. Instead, they sponsored the development of airplanes with increased range, speed, and size, forsaking the acquisition of aircraft designed or suited for cooperative missions with the army, including the cavalry. This change was intentional, although subtle at times. In the British *Army Training Memorandum no. 2 (Collective Training Period 1930 Supplementary)*, RAF orders changed their terminology from "in support of" to "on the front of" the army because the former phrase was "not suitable of application to the R.A.F."[41]

Ground troops had different priorities than aviators, as the tradeoff between mobility and speed demonstrated. High air speeds were not a measure of mobility for ground troops. Mobility had to be determined instead by where airplanes could operate. Faster planes were not always better planes. As an Air Corps colonel and instructor in the War Plans Division at the Army War College, Washington, D.C., noted, tactical aviation, particularly observation and liaison duties, required slower aircraft than those under development. Slower landing and operational speeds allowed for closer support of troops.[42] Aircraft could then take off and land near commanders and maintain aerial observation. Unfortunately for the cavalry and other ground forces, the air arm largely ignored their desire for tactical aircraft.[43]

However, there were still small positive steps. Chief of Cavalry Leon B. Kromer praised a visit to Fort Knox by students of the Air Corps Tactical School that provided the opportunity for contact between the cavalry and aviators during a demonstration of a mechanized cavalry brigade in 1937. He stated that "the more the officers of the various arms know about what others are doing, the better it is for us all." He suggested that it might also be beneficial for officers at Fort Knox to visit the depot at Wright Field because "it would no doubt prove most interesting and instructive."[44] The visit, however, did not retard the development of the independent air force theory.

Adna Chaffee, a lieutenant colonel on the General Staff, best described the frustration of cavalrymen and other army officers about the concentration on strategic bombardment over army cooperation in 1938 when he contended that "a determined army can not be shot out of position and a determined people can not be bombed into submission."[45] He maintained that air power alone could not win wars. The army needed to combine all necessary forces to win a war, and aviation was a valuable tool with which to gather important information and support ground troops. Chaffee, claiming that the use of aviation in the Spanish Civil War and in China did not contribute directly to the success of ground army objectives, warned that aviation efforts would be "largely lost" and only serve to "arouse the resentment of the rest of the world" unless aviation was used for tasks other than strategic bombardment.[46] Hawkins, also using the Spanish Civil War to criticize strategic bombing, claimed, "Airplanes have been strangely enough more effective in assisting the infantry than in bombing important centers, roads and railroads."[47]

This conclusion reflected similar findings from experiences of the British in their territories and mandated areas. In neither case did aviation acting independently of the army gain much praise from ground troops. The RAF's doctrinal transition away from tactical air support harmed the cavalry and all ground forces. With air forces more focused on independent missions, the desire expressed in earlier decades to have more integrated air components took on a more desperate tone, especially in the U.S. Cavalry. If the Air Corps and RAF would not provide aircraft to support the cavalry, horsemen would have to get their own.

Autogiros

Since the 1920s, cavalry commanders had expressed a desire for aviation components to be permanently attached to their units to improve coordination between air and ground units. Their desires were only partially fulfilled. Both the British and Americans developed tactical air support units that could be temporarily attached to ground forces, but the RAF and Air Corps had ultimate control of these elements. As a result of their provisional status, these attached aviation components did not live up to the cavalrymen's expectations. Unless aviation was assigned permanently to the cavalry division, noted a *Cavalry Journal* article, cavalry officers feared their units would have to complete tasks for which they were no longer suited.[48] Aviation had become an effective way to liaise between forward ground forces and command in the rear as well as a good collector of strategic reconnaissance and source of tactical support for ground units. A cavalry major assigned to Maxwell Field in Alabama argued that "a cavalry division has need of an organic air squadron trained to work with cavalry."[49] In the 1930s, efforts to maintain and improve ground units' aerial support became more desperate as the British and American air forces acquired aircraft and developed doctrine intended increasingly for missions other than ground support. As a result, the cavalry and other army branches attempted to find a replacement for the supporting roles that had first been appropriated and then neglected by the air force. The Office of the Chief of Cavalry sought observation squadrons to be assigned to cavalry divisions, a longstanding cavalry goal.[50]

Attempts to develop more effective tactical air support and liaisons in the United States and Britain led to trials of a newly invented type of aircraft—the autogyro, also known as a gyroplane, gyrocopter, or autogiro (the last a generic term derived from the proprietary name for the Cierva Autogiro Company's products). This vehicle appeared in the 1930s, and the cavalry believed it could be used to provide the tactical assistance that the air forces were taking away. H. F. Gregory, an army pilot assigned to test-fly autogiros, described them as airplanes with "a windmill on top."[51] They derived lift not from fixed wings but from rotors (like helicopters with the difference being that a helicopter's rotors are powered, whereas a gyroplane's are not). First-generation autogiros resembled most other small airplanes, possessing a fuselage, tail, front-mounted engine driving a propeller, and wings. The unique element was the rotor.

The autogiro flew like an airplane except that the wings autorotated. Gregory illustrated autorotation by describing how a maple seed spins as it falls to the ground. In an autogiro, an engine provides forward motion, which drives the rotors, and the rotors provide the lift.[52] This feature made them more user-friendly and less likely to crash than airplanes. If the engine stopped, gravity would pull the craft downwards, generating sufficient airflow to keep the rotors moving enough to produce lift. Fixed-wing aircraft were not good gliders when they lost power and were thus more likely to crash. Autogiros required a shorter runway than airplanes because rotors could generate lift at a slower speed than fixed wings.

This feature meant that autogiros were resistant to stalls, which occur when a lifting surface (wing or rotor) is not moving fast enough to produce enough lift to sustain flight. Even though the first autogiros had solved the problem of moving at slow speeds without the aircraft stalling, its control surfaces (the traditional wing-based ailerons, elevators, and rudder) became ineffective because they required a certain volume of air passing over them to work. Inventor Juan de la Cierva's solution to this problem was the direct control autogiro in which movements of the rotor hub, not the wings and tail, controlled the plane. The name "direct control" came from Cierva's mechanical system allowing the pilot to "achieve direct control by moving the rotor head" with a "control rod attached to the hub."[53] Historian Bruce Charnov described direct lift autogiros as the "next evolutionary stage of autogiros" and bearing "almost no resemblance to its airplane ancestor" because they no longer needed wings or tails, made useless by the new control system.[54] Cierva continued to strive to produce a vehicle that integrated collective and cyclic pitch controls and could "jump" into the air. The goal was near-vertical takeoffs that would allow autogiros to land and take off almost anywhere that had a small patch of clear ground.

Although autogiros were never able to takeoff or land vertically, they could, as Cierva claimed, fly slower and lower than most fixed-winged aircraft, requiring only small runways. They could even hover over the ground in a sufficiently strong headwind. Thus, they were, for Cierva, a better platform for reconnaissance than an airplane.[55] These characteristics of the autogiro and its ease of use made it a good candidate for use as a liaison and reconnaissance aircraft by even non–air force personnel.

Although continuingly distracted by the lure of air power independent of ground forces, the RAF did assign resources to the development of this new aircraft. The RAF assigned autogiros to Army co-operation squadrons and tested them in combined maneuvers. Because the RAF maintained control of the autogiros, British cavalrymen were not able to test the machines themselves. RAF–cavalry interactions were limited to maneuvers and confined to reconnaissance and communication. In 1933, an intercommand training exercise on Salisbury Plain tested autogiros for communication and low-height reconnaissance behind friendly lines, but the cavalry had no direct involvement.[56]

Although its service career was short, the autogiro achieved some small success in the British military. As noted by a special wire to the *Philadelphia Inquirer,* British autogiros replaced captive "sausage balloons" for artillery spotting because, unlike balloons, they were more easily camouflaged, could hover, and could fly away "with sufficient speed to offer some chance of escape from hostile aircraft."[57] By late December 1934, the War Office "decided to replace observation balloons with autogiros" due to their "obvious reconnaissance and observation potential."[58]

Additional testing with the autogiro was slow. The 1935 combined RAF and army war games produced mixed results from the six experimental Avro

Rota autogiros deployed, which did little to speed development. Difficulties with takeoffs and landings on rough terrain resulted in additional research in wind tunnels and decreased military trials.[59] The 1937 death of autogiro inventor Cierva delayed purchasing and tests further.[60] However, limited trials with direct lift autogiros continued into 1939.[61] The British army experimented with twelve C-30 wingless autogiros manufactured by A. V. Roe & Company of Manchester and had additional autogiros of various types for experimentation, research, and development.[62] The outbreak of World War II hindered additional development and ended experimentation, but Britain did employ autogiro squadrons during the war; it was the only victorious country to do so.[63] The Intelligence Division of the U.S. War Department reported that the British had two Rota autogiros in January 1940 but that experienced British observation officers were "contemptuous of this equipment."[64]

Cavalrymen in Britain did not play a role in the development of the autogiro for both organizational and practical reasons. The British experience with autogiro development reflected the larger problem that all aircraft fell under the RAF's purview. Unlike in the United States where the Air Corps and cavalry were both part of the army, in the United Kingdom the RAF controlled all aircraft development, even those intended for service alongside other branches. Although cooperation with ground units was the least desirable use of aircraft from the perspective of RAF leadership, because it weakened their arguments for their independent status, the RAF still controlled all aviation.[65]

As with other types of aviation, the Americans followed European autogiro developments. American entrepreneur Harold Pitcairn was so impressed by autogiro demonstrations in London and Paris that he joined forces with Cierva to build the first American autogiros, PC-1 and PCA-1, in 1929. Two years later, in 1931, President Hoover awarded Pitcairn the National Aeronautic Association's Collier Trophy after the PCA-1 landed on the White House lawn as part of a demonstration of its abilities. W. Wallace Kellett founded his own autogiro manufacturing company to compete with Pitcairn's Pitcairn-Cierva Autogiro Company. In the early days of American autogiro development, marketing focused on nonmilitary uses of the aircraft. Most notably, both companies sought government contracts to supply autogiros for mail delivery.[66]

In the summer of 1934, Chief of the Air Corps Major General Benjamin

Foulois was informed of both the president's interest in and the secretary of war's approval of tests of autogiros for observation at the field artillery, cavalry, and infantry schools.[67] This attention may have come from Roosevelt's knowledge of German autogiro and helicopter development.[68]

American cavalrymen wanted aircraft specifically designed to support ground forces, and if the Air Corps would not provide them, they wanted their own. The autogiro appeared to be one possible solution to their problems. Indeed, the vice president of the Autogiro Company of America (previously Pitcairn-Cierva Autogiro Company) promoted autogiros as "flying cavalry" in a confidential paper submitted to the U.S. military.[69] This claim made them attractive to American cavalrymen who wanted to continue to take advantage of aerial reconnaissance.

The American situation differed from the British one. Assigned with flying in the War Department, the U.S. Air Corps had formal responsibility for testing liaison, reconnaissance, and tactical aviation, yet the Air Corps was prepared to allow other parts of the army to test new technologies that might fit their particular tactical support requirements.[70] Lieutenant Colonel Harold E. Hartney, the aviation technical adviser to the Senate Commerce Committee, testified to the House Military Affairs Committee in 1938 that the Air Corps had "awful[ly] big problems of their own to solve at the present time" and was not concerned if other branches experimented with new types of tactical aircraft.[71]

The autogiro became the focus of American cavalrymen's attempts to gain control over tactical air assets from the "newly emboldened Air Corps, which was," according to Cierva, "willing to sacrifice the development of tactical capabilities in support of its greater strategic visions of airpower doctrine."[72] The viability of the autogiro for American forces "played out against . . . [a] backdrop of intra-service rivalry and shifting doctrine" as army branches, including the Air Corps and the cavalry, were competing for predominance and money.[73] Although Holley argued that the "autogiro [was] looked upon by some authorities as possibly being the replacement for the horse for reconnaissance purposes,"[74] some of the most vocal support for the autogiro came from the American cavalry.

Far from rejecting technological changes, the U.S. Cavalry actively campaigned for the autogiro. In the mid- to late 1930s, senior cavalrymen became increasingly interested in promoting autogiro procurement and experimentation. Beginning in July 1935, the Office of the Chief of Cavalry requested an autogiro to experiment with at Fort Riley.[75] Unsatisfied with the pace of delivery, less than a month later, Chief of Cavalry Major General Leon B. Kromer asked the Office of Chief of Air Corps to "hasten procurement of an autogiro" for cavalry experimentation so he could determine if it would be "of material benefit to the cavalry arm in carrying out its assigned combat roles."[76] The desire for testing did not end with the chief of cavalry's office. The colonel commanding the 1st Cavalry requested two autogiros for testing with the 7th Cavalry Brigade (Mechanized) at Fort Knox, Kentucky, in September 1936.[77]

The cavalry's interest and the correspondence did not relent. Only three months after its first letter, the Office of the Chief of Cavalry sent another letter to the adjutant general to emphasize that the chief of cavalry was still "intensely interested" in the development of the autogiro and would like an autogiro at the Cavalry School in May 1937 and then for testing with the 7th Cavalry.[78] The office labeled the "experimentation and development of the autogiro as an instrument of command and aerial observation" as "a high priority project" and continued to try to get autogiros for testing.[79] In March 1937, the commandant of the Cavalry School, adding his name to the list of cavalry officers who expressed a desire to work with autogiros, asked to test autogiros at the Cavalry School.[80]

Finally, after a delay caused by technical and safety difficulties, the Cavalry School at Fort Riley received an autogiro in May 1937 and began testing its capabilities. Although the Army Air Corps was willing to share development of this new technology with other agencies, it wanted to determine its safety first. The Air Corps did not receive its first autogiro, the Kellett YG-1, until October 1936.[81] Delivery of autogiros to the cavalry had to be delayed because one was in the factory for repairs from April to July 1937 and another disintegrated in the air in June 1937. The Air Corps even suspended their own tests in June 1937 until, as an air corps officer stated, the "necessary engineering changes" were made to "preclude danger of another failure."[82]

Despite these complications, tests with the cavalry began on May 10 and lasted until May 28 to assess the suitability of the YG-1, a wingless direct control autogiro, built by the Philadelphia-based Kellett Autogiro Corporation, and to compare it to airplanes.[83] The cavalry wanted to know if the autogiro could enable commanders to "observe the enemy situation in the immediate vicinity of [their] command and direct from the air if necessary the movements of [their] own troops."[84] The air speed of the autogiro ranged from 16 to 130 miles per hour, with a cruising speed of 105 miles per hour and a maximum diving speed of about 140 miles per hour. Its maximum ceiling was eighteen thousand feet and its minimum height was whatever was needed to clear any obstacle safely. The weight-carrying capacity was approximately 500 pounds, which usually consisted of a pilot, a passenger, and thirty pounds of baggage. The lower the weight, the less distance it needed for takeoff.[85] Although there were no direct comparison trials between autogiros and airplanes at the Cavalry School, the testing concluded that airplanes surpassed autogiros in strategic reconnaissance and column control, battle reconnaissance, contact missions, and artillery adjustments. These advantages were due to the autogiro's slow top speed, small carrying capacity, short cruising range, and lack of defensive armament. The autogiro's slow minimum speed and its ability to land and take off in small areas proved superior to the airplane with regard to command reconnaissance (especially when conducting terrain studies), night reconnaissance, and command control (defined as the transmission of orders and messages throughout the command).

The Cavalry Board concluded that an autogiro and Air Corps pilots "should be attached to each regiment of cavalry and each regiment or separate battalion of field artillery serving with cavalry." In addition, the board recommended obtaining additional autogiros for the Cavalry School, the 1st Cavalry Division, and the 7th Cavalry Brigade (Mechanized) to improve the "technique of its employment." From Fort Riley, the Kellett YG-1 Autogiro was sent for additional testing at Fort Knox, Kentucky, with the 7th Cavalry.[86]

These opportunities to test autogiros did not satisfy the cavalry. Testifying in April 1938 before the House Committee on Military Affairs on the authorization of funds for autogiro development, Colonel Guy Kent from the Office of the Chief of Cavalry asked to continue to test the autogiro to determine all of its abilities. Even though the chief of cavalry was not completely satisfied with the thoroughness of previous testing, he was convinced that "an autogiro . . . should be attached to each regiment of cavalry."[87] His office wanted additional opportunities to work with autogiros to determine if they would allow a commander or staff officer to observe and command from the air.[88] The committee appropriated two million dollars to be "expended for the purpose of Autogiro research, development, and procurement for experimental purposes."[89]

The cavalry's request to test the autogiro further was approved, and additional tests began at Fort Riley in August 1938.[90] Cavalry commanders continued to develop this unique aerial craft without the intermediary of the army air service command structure. Experiments tested the possibilities of autogiros improving communication and observation with both horsed and mechanized units. Determining the autogiro's feasibility to complete a myriad of missions entailed familiarization flights, night landings, road reconnaissance, command missions, and various other tests.

The chief of cavalry was not wed to a particular technology but to fulfilling the cavalry's needs. He was interested in the development of autogiros and willing to embrace any new cost-effective and useful technologies that shared "the best qualities of the autogiro."[91] In August 1938, the chief of cavalry told the adjutant general that the cavalry remained "intensely interested in the development of the autogiro or a similar type of rotary wing aircraft for command and other purposes with both horse and mechanized cavalry."[92]

The 1938–39 trials produced only mixed support for autogiros. Like previous aircraft in the early stages of development, the autogiro had its problems, including low carrying capacity, vulnerability to ground fire, and structural deficiencies. These limitations did not prevent Colonel Charles Burnett from sharing the cavalry's excitement in a letter to the Kellett Autogiro Corporation in August 1939: "We recognize that it has great possibilities for use in the Cavalry Service and we are hopeful that in the future it may develop even greater possibilities than at present."[93]

Unfortunately for the Kellett Corporation, that enthusiasm was not universally shared. In late 1939, a board was formed at Fort Knox, Kentucky, at the behest of the president to test and compare the YG-1 B autogiro and the O-47 observation plane to determine if autogiros were worth the investment. The board consisted of two cavalry officers, Colonel Jack W. Heard of the 13th Cavalry and Lieutenant Colonel Victor W. B. Wales of the 1st Cavalry; a field artillery captain; and two pilots, both first lieutenants, one from an observation squadron and the other from the air corps.[94] In early 1940, the board reported that the autogiro was superior to the O-47 in eight of the mission categories and about the same in five additional tasks, but the autogiro's higher cost and limited availability were major drawbacks. The board's report depended heavily on research ordered by the committee and completed by Air Corps 1st Lieutenant Robert M. Lee. He provided informa-

TABLE 2

Relative Cost and Estimated Performance of the O-49 versus the Autogiro

	O-49	AUTOGIRO
Cost (based on number procured)	For 5: $43,850.08 ($8,770.02 each) For 10: $37,473.67 ($3,747.37 each)	For 7: $34,782.42 ($4,968.92 each)
Highest speed	137 mph	150 mph
Take off and clear obstacle 50 ft.	350 ft.	250 ft. (no roll)
Landing and clear obstacle 50 ft.	350 ft.	250 ft. (no roll)
Minimum speed	29.5 mph	20 mph
Ceiling	20,000 ft.	20,000 ft.
Availability of spare parts	Same as fixed wing	Limited
Availability of raw materials	Same as fixed wing	Limited

tion about production costs as well as the availability of spare parts and raw materials. Lee made a slight alteration to the assignment by replacing the O-49 for the O-47 for comparison with the autogiro because, he argued, it "more nearly correspond[s] to the autogyro." Table 2 reports his findings.[95]

Using Lee's findings, the board concluded in rather strong language that the autogiro should not be added to the Observation Squadron attached to the cavalry, instead recommending the development of "a light, slow airplane" to perform essential missions.[96] The Cavalry Board recommended focusing on developing light conventional airplanes to operate with the cavalry on observation, command, and liaison missions. Although the board members had yet to see such a plane, they believed there was one in development.[97] In 1941, the army acquired the Taylorcraft L-2, Aeronca L-3, and Piper L-4, all of which were civil aircraft modified for military use.

Three years earlier, the chief of cavalry had already given his support to testing low-speed conventional aircraft for observation, liaison, and command purposes in addition to the autogiro. One such aircraft that he suggested, the B-42, proposed by the Crouch-Bolas Aircraft Corporation, was a "low-wing, cantilever monoplane" powered by four engines.[98] This plane did not end up in production.[99]

Despite the Cavalry Board's suggestions to focus on airplanes, the cavalry did not entirely abandon the autogiro, continuing to experiment with it in the 1940s. Autogiros were tested during the 1940 army maneuvers with the 1st Cavalry. The board's recommendations were not ignored but modified. As Cavalry Lieutenant Colonel Willis D. Crittenberger wrote after flying the autogiro, "I am for the autogiro until something better is actually in our hands—not promised." He thought the autogiro almost filled the cavalry requirements and was better than any other substitutes in existence and should be used.[100]

Despite its potential, the autogiro was a failed technology in the United States because it was never widely adopted by civilian or military organizations. It did not become a permanent part of the American cavalry, or the army in general, because of financial and operational considerations. Another hindrance to autogiro development was the outbreak of World War II, because design developments had to focus on aircraft immediately available for action. Nevertheless, the cavalry's intensive testing of the experimental aircraft demonstrated a desire to develop and deploy an aerial tool alongside its mounted ground forces.

The New Challenger—Tanks

While cavalrymen actively experimented with aviation, mechanization in the form of tanks and motor cars replaced aviation as the new technological challenge to the cavalry's roles in the early 1930s.[101] Mechanization's advocates circulated familiar predictions about the ability of their new technologies to replace the horse cavalry, while cavalrymen cited aviation to show how, yet again, the technological promoters were mistaken and armored and mechanized vehicles would not necessarily make the horse cavalry obsolete.[102] A cavalry lieutenant colonel made this connection clear when he argued that "theories of the extremists in mechanization are not likely to have more effect on our doctrine than those of certain extremists in aviation."[103] Cavalrymen utilized a similar framework to respond to this new contender as they had to aviation.

First, they attacked overconfidence in new unproven technologies. Although it was "natural that new arms and armaments should arouse ardent enthusiasm and acquire stanch supporters," untested innovations usually fell "short of expectations."[104] Major George S. Patton Jr. agreed with this observation from a foreign article republished in the *Cavalry Journal*, attacking the idea that motorized and mechanized vehicles would destroy the cavalry by referring to the inaccuracy of earlier predictions that airplanes would replace the cavalry. He stated "experience of the effects of storms, fogs, darkness, forests, and enemy planes has so modified this view that now the airplane is considered as the ally not the supplanter of cavalry for strategic reconnaissance."[105] Patton provided additional examples of aviation's limitations, including the inability to obtain identifications and to maintain constant surveillance, as evidence that such predictions could not be trusted.[106] Patton, author of many *Cavalry Journal* articles in the early 1930s, compared the chariot, elephant, gunpowder, dynamite gun, and submarine to the tank, airplane, and gas, noting that each had at one time or another been "acclaimed as the mistress of the battlefield" and adding that none lived up to these claims.[107] Patton argued that the "fraternity of motorists" and "gasoline neophytes" was leading a "mechanically minded and gullible public" astray.[108]

Before the military accepted a new technology, cavalrymen argued it must be tested, the same contention they had made about airplanes throughout the 1910s and 1920s. One American cavalry colonel proclaimed that his colleagues were "earnestly striving to determine all the possibilities of these new developments in the interest of National Defense and to fit each into its proper place in our respective combat teams." Rather than blindly adopting new technologies, he thought it more prudent to test each new invention to make sure it worked before discarding "the experiences of centuries in equipment to put our faith in a new 'gadget.'"[109]

At the beginning of the decade, the cavalry's agenda to modernize itself was evident in the newly created "Progress and Discussion" section in the *Cavalry Journal.* Introduced in January 1930, this section's purpose was to record the "state of development of various items of material in which the cavalry is particularly interested, progress in technique or tactics" and "to encourage discussion of matters of general professional interest to cavalry officers." The introduction invited "suggestions concerning the new methods developed and discussion or constructive criticisms of present methods, equipment, etc."[110] Throughout the 1930s, this new section discussed evolving technologies such as armored cars, machine guns, and semiautomatic rifles. A 1935 article in the *Cavalry Journal* stated that "the suggested changes in organization and equipment that now come into the Chief of Cavalry's office and appear in print are conclusive proof that the cavalry as a whole is wide-awake to new developments for this arm."[111]

Testing the possibilities and use of armored cars and tanks in training maneuvers in the 1930s mirrored those of the 1920s with aircraft. As before, cavalrymen argued that the maneuvers using these vehicles showed the importance of cooperation between horsed and mechanized units, not that the latter replaced the former. As before with aviation, the difference between cavalrymen and the new technology supporters was a matter of degree. The cavalry did not reject mechanized vehicles outright, but they actively attempted to evaluate this new technology, even improvising their own armored cars when the army provided none.[112]

Reports on maneuvers describing the problems of the new wheeled and tracked vehicles in the 1930s closely resembled the arguments about aviation's limitations in the 1920s. The 1929 American cavalry division maneuvers "marked the debut of the armored car in maneuvers" with the cavalry.[113] Terrain, weather, supplies, and an inability to maintain continuous contact all harkened back to early objections to aviation. For mechanized vehicles, the inability to work in various types of terrain proved to be their major limitation.[114] Throughout the decade "roads, bridges, forests and mountains still oppose[d] mass employment of motor vehicles."[115] Exercises during the rainy season revealed that "mechanized cavalry, tanks and truck trains encountered considerable difficulty except on paved roads."[116] Nor could mechanized vehicles "maintain continuous contact irrespective of visibility or weather conditions."[117] Additional shortcomings included the dependence of armored forces on fuel supplies and spare parts for field repairs.[118]

The British *Cavalry Journal* claimed that horse cavalry was still needed because mechanized vehicles made attractive targets for enemy fire. Therefore the "obligations of reconnaissance and 'screening,' which are the essence of dispersion, can never be adequately carried out by the man in the armored vehicle, since, in his steel box, like the crab in its shell, he exhibits a perfect example of high-vulnerable concentration."[119] Massachusetts State Representative Henry Cabot Lodge Jr., a 1st lieutenant in the cavalry reserve, also rejected the claims made by mechanization's supporters including Captain B. H. Liddell Hart's assertation that the cavalry was proven obsolete "by 'assessing the issue mathematically.'" Lodge argued that Captain Hart's dismissal of the cavalry, based on his comparison of the speeds of man on foot, on horse, and in motorized vehicles, was too simplistic. Lodge's tour of

active duty on the Mexican border convinced him that the "manifold variet-
ies of terrain" an army may have to traverse made all three "methods of locomotion" necessary.[120] The American *Cavalry Journal* reprinted an editorial
that originally appeared in a Milwaukee newspaper in which it was noted
that observations such as these should convince skeptics "of the importance
of *real* cavalry in any modern scheme of war."[121] The assertion here was the
true cavalry was horsed, not wheeled or tracked.

The belief that mechanization was not yet a mature enough technology
to replace the cavalry appeared in cavalry journals. An article in the U.S.
Cavalry Journal noted the influence of the motor-minded public in discussions of the replacement of horses by motorized vehicles but remarked,
"While it is conceivable that his hoof print will someday be found only in
the fossilized exhibits of the museum, the tire tread cannot as yet claim
mastery over his domain."[122] This argument was just like previous statements made throughout the preceding two decades that although aviation
may one day replace the cavalry in reconnaissance, that day had not arrived.

An American infantry lieutenant colonel rejected the argument that the
new inventions of airplanes and scout cars made the cavalry lose its label
as the "'eyes and ears' of armies," a description it had held for centuries. He
admitted airplanes and scout cars may be the eyes of the army, but "it will
be a long-time before they become the ears of armies . . . and there will be
situations" like bad weather "where neither the one nor the other will even
be the eyes."[123]

Cavalryman argued that mechanization and motorization, like aviation,
would strengthen but not replace horse cavalry. One of the most elegantly
constructed arguments praising mechanization came from a discussion
in the American *Cavalry Journal* about foreign views of mechanization. It
stated that "far from being the condemnation of the cavalry, the motorized
engines constitute for it the opportunity for a resurrection, in which, with
formidable means at its disposal, the cavalry spirit will be able to amplify in-
finitely its innate qualities of boldness, skill, and heroism."[124] Lieutenant Colonel Adna Chaffee, a member of the General Staff, stated that both cavalry
and mechanized vehicles would "be necessary" because they "supplement
each other where combat [is] moving even as the cavalry and airplane do
today in the duty of reconnaissance."[125]

Throughout the 1930s, training reports, student papers, and articles in the *Cavalry Journal* (both in the United States and United Kingdom) argued that mechanization would help the cavalry. A report of the 1931 British collective training period suggested that scout cars were valuable and saved horseflesh.[126] Major J. T. Pierce's 1932 Command and General Staff School student research paper stated that mechanized forces do "not displace cavalry, nor can they be used interchangeably, its proper place is an adjunct to cavalry and so used will broaden the scope of cavalry and increase its combat efficiency."[127] Captain William R. Irvin in a similar paper two years later concluded that the "advent of the motorized vehicle and armored car [had] greatly increased the mobility and operating range of cavalry in combined reconnaissance."[128] Citing the combined brigade maneuver of the 8th Brigade with the 6th Cavalry in May 1935, two cavalry officers, a lieutenant colonel, and a captain argued that scout cars increased "greatly the efficiency of the cavalry, extending its vision . . . by their information, largely negative, they lessen the worry over security." Although they "take an enormous load of work off the horses," neither scout cars nor "anything resembling them" could replace mounted troops because motors are only faster in "exceptional terrain" and no machine in existence could "do everything a horse can do."[129] In reconnaissance, *Tactics and Technique of Cavalry*, an American text used for cavalry training in 1937, stated that airplanes, motor trucks, scout cars, armored cars, and combat cars added to large units "greatly augment[ing] its powers of reconnaissance" and had "materially added to the mobility of the cavalry and increased its range."[130] *Tactics* concluded that despite this value, mechanization still did not "replace any other combatant arm, but tremendously assist[ed] the operation of both infantry and horse cavalry."[131]

Hamilton Hawkins thought that tank units could "be of great assistance in special situations" and "may be a valuable adjunct" to the cavalry.[132] A mechanized vehicle could "probably render much more important services if it is operated in combination with horse cavalry than it could alone." Vehicles could supplement the cavalry when they performed "covering force duty," including "reconnaissance, screening and security . . . delaying actions, exploiting a success, combination with infantry in battle, holding positions until infantry [could] arrive or until it [could] withdraw, and in the pursuit."[133] This desire for a complementary horse and mechanized cavalry was made clear in 1939 when Chief of Cavalry Major General John K. Herr told the House Subcommittee on Military Affairs "that combined with a proper proportion of mechanized cavalry, the capabilities of cavalry are greatly increased."[134] The horse could not be fully replaced by any combination of vehicles, particularly for tactical reconnaissance in wooded areas.[135] Herr concluded by saying, "It is my fixed opinion that, although in some cavalry missions it may be better to use horse cavalry alone or mechanized cavalry alone, on the whole the best results can be accomplished by using them together."[136]

Despite the many similarities between how cavalrymen responded to aviation and to mechanization, significant differences emerged. Unlike the previous debates on aviation, the discussion about mechanization focused primarily on eliminating the horse component of cavalry and not on whether the cavalry branch was obsolete. The arguments and terms in this decade increasingly pitted animals against machines. Machines, capable of moving more quickly than horses and armored for protection against modern weapons, posed a different challenge to the traditional cavalry. Despite his article's title, "Exit the Cavalry . . . Enter the Tanks," Army Chief of Staff Douglas MacArthur did not proclaim the cavalry's end in 1931, but he predicted that in situations where the "country [was] not too rough cavalrymen of the future [would] ride . . . gasoline steeds into battle."[137] The annual British training memoranda following the collective and individual training seasons reflected this change in the standing of the traditional cavalry. By 1931, the cavalry no longer had a separate section in the reports of the training periods, and a tank section appeared in 1932.[138]

Evaluations and reports of maneuvers provided support for horses in their competition with machines. After reviewing the maneuvers at Fort Benning in May 1935, two cavalry officers argued, "There is no machine yet invented that can do everything a horse can do."[139] British collective training maneuvers in 1937 experienced problems with the abilities of machines to operate in certain terrains that were not a problem for horses.[140] In fact, Colonel Bruce Palmer, commander of the 1st Cavalry Mechanized at Fort Knox, Kentucky, maintained that "the horse is more indispensable to the army today than . . . ever" because the "versatility of the horses is the key to the whole matter, these troops can maneuver on any terrain, day or night regardless of weather conditions."[141] The reporter who recorded this statement noted that the colonel's explanation would be of "interest to many civilians who [had] been unable to understand the value of horses in modern warfare."[142]

Another difference between the cavalry–aviation debate and the cavalry–mechanization debate was that the cavalry succeeded in gaining at least temporary control of the mechanization process. While failing in the 1920s and 1930s to gain its own permanent air component, the cavalry succeeded in gaining its own motorized and mechanized vehicles in the 1930s and was transformed in the process. After the 1931 collective training period, the British Army adopted one commander's recommendation that "an Austin Scout troop of 5 cars should be included in each divisional cavalry regiment in peace time."[143] A committee considering the organization of the mechanized cavalry and the Royal Tanks Corps reported that the "development of motor transport in all its forms has made it necessary to equip the cavalry arm with fast moving armoured vehicles to enable it to perform its traditional role."[144] A lieutenant-colonel believed that some of the problems of working together to determine the appropriate employment of horse and mechanized cavalry would be solved by working together as "part of the same arm" without "antagonistic prejudice."[145] This combination was clearly demonstrated when the British *Cavalry Journal* added the subtitle *Horsed and Mechanized* in 1938.

During the 1930s, the cavalry's response to mechanization pitted those cavalrymen advocating complete mechanization against their more cautious peers who wanted to retain some horse units, which included several high-ranking officers. General Hugh A. Drum's experiences and observations of the terrain in China and Spain and during the 1938 command post exercise caused him to conclude that the "proposal that motors replace all animals [was] an extreme view not warranted."[146] In a 1939 congressional budget hearing, General Malin Craig defended the continued existence of horse cavalry by arguing that "there are many circumstances where it is essential and where mechanized cavalry cannot take its place."[147] As early as 1930, British cavalryman Lieutenant-Colonel E. G. Hume also argued that the public belief that "cavalry should be composed entirely of horsed troops, or be completely mechanized" was unsound and that further examination should be made of the "respective spheres of usefulness and limitations of these two methods of mobility: the machine and the horse." They should cooperate "wholeheartedly."[148]

The division between horse cavalrymen and mechanized cavalrymen became so pronounced that some individuals such as Major R. W. Grow tried to heal the wounds. He remarked that in his "humble opinion it is high time to drop all this controversy between horse and mechanized and get together as *cavalrymen*." He maintained "to be pointed out as a 'horse' cavalryman savors too much of hidebound tradition," while "to be pointed out as a 'mechanized' cavalryman savors too much of a scatter-brained enthusiast without his feet on the ground." It was his belief that "neither the four-footed horse nor the steel horse, in themselves, make cavalry." He argued that they all should be horsemen but "above all . . . cavalrymen."[149] A contributor to the British *Cavalry Journal*, identifying himself only by the initials A. F. V., also noted the tensions between tank men and horse men. He acknowledged that the division, although overemphasized by outsiders, did exist. He maintained that it seemed "to be forgotten by the disputants that the horsemen, and the tank men . . . have to fight in any future war besides, and not against each other." They all had the same goal: achieving victory. This desire for victory, he argued, "must enable us, if we have not lost all sense of reason, to hope for, and even rejoice in, the success of the rival arm over our own."[150]

The American chiefs of cavalry attempted to be a progressive and unifying voice during this debate. Both Major General Kromer and Major General Herr desired further mechanization. Kromer, chief of cavalry from 1934–38, "recognized from the start that the 'iron horse' [had] opened to cavalry a greater sphere, and that cavalry must unhesitatingly seize and exploit to the fullest extent consistent with its development, an implement that bids fair to greatly enhance the powers of the arm."[151] Unlike many of his more traditional contemporaries, Kromer "steadfastly held . . . that mechanized cavalry [was] cavalry."[152] The 1937 curriculum at the Air Corps Tactical School also expressed this viewpoint. In these courses the term cavalry referred to both armored cars and horses. One lecture stated that cavalry could use "the *machine*, the *horse*, or *both*" to achieve its strategical or tactical mobility.[153]

Herr, chief of cavalry from 1938–42, also considered mechanization an "integral part of each arm" and not separate.[154] In 1938, he reminded his fellow cavalrymen of the importance of the cavalry's mobility, which he maintained was centered on the horse, but that horse could be the "iron horse or the horse of flesh and blood."[155] Herr demonstrated that cavalrymen could embrace mechanization yet still maintain horse elements. Herr was convinced by eight years of study and development by the U.S. Cavalry that mechanization could help in the completion of cavalry missions "to a very considerable extent"; however, it still could not fulfill all of the requirements of mobile warfare. The cavalry must include both mechanized vehicles and horses; if not, it would be "courting disaster."[156] There needed to be a proper balance between horse and mechanized units.[157] *Tactics and Technique of Cavalry* conveyed a similar opinion: "Although there are many means of transportation more rapid than the horse, none of them can completely replace him . . . mud, snow, and shell torn roads still hold their terrors for the motor vehicle, while fog and low visibility often render impotent the best efforts of the air force . . . the horse can still work efficiently under these trying conditions."[158]

Arguing for British Cavalry Reductions

In the 1930s, mechanization increasingly replaced aviation as a justification for the reduction of cavalry. Cavalrymen and supporters of the continuation of horse cavalry employed similar arguments to those they had used to defend their branch when confronted by aviation. These contentions included claims that mechanization was not advanced enough to replace the horse, tanks could not accomplish what their proponents predicted, mechanization actually strengthened the horse cavalry, and those who contended that the Great War proved the cavalry was obsolete misunderstood the war. Historian Edward Katzenbach observed that during the mid- to late 1930s, when horse cavalry was most threatened by military reorganization plans, the value of the horse in the Great War was emphasized more than in any time previous.[159]

Parliamentary debates in the 1930s provide excellent examples of the divisions between politicians and military officers over the relative values of mechanization and the cavalry and the future of both. Member of Parliament John Joseph Tinker took the role of mechanization advocate and cavalry detractor. In 1932, he targeted the cavalry for reduction during a discussion over army estimates.[160] Tinker argued that mechanization of the army was "for the purpose of removing what are called obsolete units, and in these days cavalry cannot be called a useful arm of the service."[161] Tinker brushed aside accusations that he failed to consider and appreciate the battle honors afforded those regiments, stating "every one of us who fought in the War attained some honour for trying to save the country."[162] He campaigned to eliminate the cavalry, regretting that "we have lost many thousands and millions of pounds by our delay in this matter."[163]

Against such adamant opponents, the cavalry needed all of the supporters it could get. One appeared from an unexpected quarter: the RAF. Wing Commander Archibald James stated that Tinker appeared to be putting too much value on what had occurred on the Western Front. James argued that although he was a member of the RAF, a service that did not dispose him to favor the cavalry, he recognized the nation's responsibilities included "very 'unmechanisable' parts of the world," such as Poland, and that it was not yet "proved that our types of mechanised machines in substitution for horses are satisfactory."[164] Mechanization was coming, he agreed, but "you cannot hurry these things." Brigadier-General Sir Henry Page Croft concurred, citing conditions in Iraq and proclaiming that the "total abolition of the cavalry would be a frightful blunder."[165] He warned that once horses were eliminated, it would be difficult to bring them back due to the time necessary to raise and train new mounts.

Despite these warnings, other military officers disagreed with their colleagues and echoed Tinker's arguments for increased economy by reducing or eliminating the horse cavalry and replacing it with mechanization. Although Major James Milner did not state directly that tanks would supplant cavalry, he argued that the "cavalry have been obsolete on the Continent of Europe since the Battles of Crecy and Poitiers" leaving "ample scope for a great reduction in the cavalry arm," which would free money for the acquisition of tanks.[166]

Politician Allen Bathurst, Lord Apsley, who had served with the Royal Gloucestershire Hussars during the Great War, came to the cavalry's defense. He called for an increase in the cavalry establishment to correct the handicap experienced by field officers who did not have enough cavalry in the Great War. He supported Wing Commander James's assertion that horse cavalry was needed to "work in country where mechanical vehicles cannot proceed, and . . . to carry out reconnaissance . . ." but before he could finish his statement, he was interrupted by Milner who stated simply "do it by aeroplane."[167] Lord Apsley responded, "Tactical reconnaissance by air is useless. Strategical reconnaissance is not so bad, but it is impossible to tell from the air where the flank of an advance is, where troops are deployed, what form the troops take and what positions they are holding. You can only get tactical reconnaissance by drawing fire. If you are going to retain any cavalry it should be brought up to strength and made efficient." Apsley was not against technological change. He suggested that "as many as possible of the motorized cavalry should be equipped with light tanks" and that the cavalry should be given control of its own aviation in the form of light airplanes for reconnaissance.[168] He was primarily concerned with the cavalry being able to accomplish its job—reconnaissance, rearguard actions, and pursuit—and not with the means. "We want the best, quickest and most effective means of transport to get these functions performed," he argued. Apsley advocated training "plenty of men" to use whatever type of mount was available, whether it be horse, mule, camel, donkey, motor car, light tank, or aircraft. He also hoped that horses from recently mechanized units in Egypt would not be given away or killed but given to the remaining horse cavalry regiments.[169]

Stereotypes

The defense of the horse had been part of the debate over the replacement role of the bicycle, motorcycle, and airplane, but in the 1930s, this defense became more common and more central for cavalrymen. The argument that the man (and the horse), not machines won wars, so common in the military and public in the 1920s, dropped in visibility by the 1930s except for in the writings of cavalrymen. Cavalrymen during this period were more susceptible to the charges of "muddy-boots fundamentalism" and irrational and unquestioning support of the horse that historian Brian Linn described as part of the negative connotations of the "Heroes" intellectual tradition of the United States Army. [170] American cavalrymen tended to "characteristically dismiss those who [sought] to impose predictability and order" onto war, which they viewed as "chaotic, violent, and emotional."[171] More cavalrymen slipped into emotional posturing, elitism, and grandiose proclamations.[172] This defense of the traditional horse cavalry was atypical of arguments during the first twenty years of military aviation, but it was a major component of the response to the threat of mechanized ground vehicles, especially in Britain.

Two British poems clearly demonstrated the emotionalism of cavalrymen fearful of losing their horses to mechanization. Significantly, they appeared in 1938 and 1939, after mechanization had clearly supplanted the horse in Britain. These poems, published in the British *Cavalry Journal: Horsed and Mechanized,* eulogized the rapidly disappearing army horse. In one, "Night Guard," the unnamed author recalled the sense of familiarity and comfort when their steeds remained.

It was different when we had horses;
The stables were warm and snug,
You could go and chat to your "long-faced pal"
And straighten his rucked-up rug.
There were munching sounds in the darkness
As you opened the stable door,
And the rattle of chain on head stall,
Or the clink of a hoof on the floor.

He contrasted these happy memories with the present in his next stanza, noting the characteristics of the "garage block":

But now there's only a garage block,
Dark and lifeless and dumb,
And you're bored quite stiff, and you wonder if
Your relief will ever come.
You rattle a door or window,
And continue to the weary round;
You gaze at the clock or whistle a tune,
Just for the sake of the sound

The author ended his poem on a more positive note, attempting to raise the spirits of his fellow cavalrymen by stating

And there's one very big advantage;
When the leave seasons come on
You can turn the key on your tanks and trucks,
And pack up your kit and be gone.
Now that means a hell of a lot chaps,
So you needn't moan or take fright;
The army's seen changes before, pals
It'll weather this lot alright.[173]

The author of the second poem, Major Cyril Stacey, formerly of the 14th King's Hussars, was far less willing to embrace mechanization and motorization. He praised the old ways and wondered if horses might be needed again.

No more the dung pit's reek perfumes the breeze,
No more the squadron-leader shouts out "march at ease,"
No more the troop horse searches in his manger
Oblivious to any thought of danger.

He also observed the good fortune of those in a few special regiments who retained their horses:

> Gone are those chestnuts, browns and bays
> From all except those lucky 'Royal' and 'Greys';
> And Household Cavalry, who when at home
> Still use the hoof-pick and curry comb.
> Hussars, dragoons, and lancers still bear ancient names,
> The men content to play dismounted games
> While tanks and lorries grimly thunder past,
> Replacing those dear gees that have been cast.
> Of petrol now they draw a daily ration
> Instead of oats and hay which were in fashion.
> Gone is the farrier and the skillful vet.;
> Who know but we may want them badly yet?[174]

Despite these emotional arguments, every defense of the horse was not unthinking nostalgic conservatism, although much of it could be construed that way. Cavalry historian Katzenbach has observed that with the exception of these poems, the cavalrymen's arguments in their various professional forums were "absolutely sound." Tanks and airplanes could not replace the horse "until such time as it could perform all the missions of the horse."[175] The major issue overlooked, however, was whether these missions were valuable.

The contention that the cavalry and its leaders were backward and conservative took a powerful visual form in Britain on April 21, 1934, when David Low, the famed cartoonist for London's *Evening Standard*, created the "powerful icon" of "reactionary stupidity" in the British military: Colonel Blimp. It is not an accident that this pretentious yet inept character was identified as a British cavalry officer.[176] Cartoons fearing the colonel proved immensely popular, so much so that his name entered the language. The *Oxford English Dictionary* identifies Colonel Blimp as a character "representing a pompous, obese, elderly figure popularly interpreted as a type of diehard or reactionary."[177] Low intended, however, to typify the "current disposition to mixed up thinking, to having it both ways, to dogmatic doubleness, to paradox and plain self-contradiction."[178] Blimp was a fitting name, as it referred to a type of airship—a balloon, a gas bag. Blimp initially took the form of a cavalryman because Low had read a letter from an officer that the "cavalry should continue to wear their traditional uniform and spurs even when they were mechanised."[179] The defining characteristic of Blimp, according to Low, was his daftness. He strongly emphasized that Blimp did "NOT represent a coherent reactionary outlook so much as slapdash stupidity."[180] Low used Blimp not to criticize the military but political issues and individuals. Despite his attempts to protect his definition of his character, Low quickly lost control of his creation. Low's readers turned his comic figure typifying stupidity into a symbol of the conservatism of British military officers, a matter of great concern to the British public. Yet most Britons did not exclusively associate Blimp with the cavalry but instead perceived him as a generic British army officer.[181]

C. S. Forester conveyed a similar message with his 1936 novel *The General*, which strengthened the popular belief that cavalrymen and commanding officers were backward and reactionary. The main character, Herbert Curzon, was Forester's "caricature of the best and worst" of British commanders during the Great War, "hide bound, traditional and utterly devoid of imagination, yet, brave and honorable to a fault."[182] Curzon, unsurprisingly, was a cavalryman who distrusted the highly educated and military theorists. He believed that any man discussing the theory of war would almost surely "bring forward some idiotic suggestion, to the effect that cavalry had had its day and that dismounted action was all that could be expected of it, or that machine guns and barbed wire had wrought a fundamental change in tactics, or even—wildest lunacy of all—that these rattletrap aeroplanes were going to be of some military value in the next war."[183]

Curzon was mortified that one of the "feather-brained" subalterns voluntarily and with enthusiasm "quitted the ranks of the twenty-second lancers, the Duke of Suffolk's Own, to serve in the Royal Flying Corps."[184] Curzon complained that this subaltern had further insulted the cavalry by having "had the infernal impudence to suggest to the senior major of his regiment . . . who had won the Battle of Volkslaagte by a cavalry charge that the time was at hand when aeroplane reconnaissance would usurp the last useful function which could be performed by cavalry."[185]

In reality, of course, the majority of military officers in Britain and the United States were not Blimps or Curzons. There may have been some who agreed with the poems cited above and desired to charge their horse once more at the enemy when defeat appeared certain, as Curzon did in the last pages of Forester's novel, but to characterize all cavalrymen who defended the continued utility of the cavalry and horses as hidebound anti-intellectuals is to do a disservice to the great number of cavalrymen who pragmatically addressed new technologies.

Hamilton Hawkins, labeled by scholar Alexander Bielakowski as traditional,[186] rejected the contention that opposition to mechanization came from "ultra-conservatives" and "old fogies" in the cavalry, claiming that airplanes and mechanized vehicles realistically could not fully replace the cavalry. Instead, "brilliant success" could occur in battle if an air force and mechanized force combined with the motor transport of supplies to cooperate with the cavalry.[187]

Although some cavalrymen may have been overzealous in their defense of the horse, the cavalries of the United States and Britain examined and attempted to incorporate new technologies. Presenting a less nostalgic view of this involuntary "great metamorphosis" that the cavalry was experiencing, a 1938 British *Cavalry Journal* editorial argued that while "a person may or may not like change" it was "apparently ... quite inevitable."[188] Consequently, the journal shifted coverage from saddles and sore backs to pistons and differential gears. Despite this change, the editor argued that cavalrymen may have switched their mounts but would continue to complete their duties with the same devotion as before.[189]

The overly sentimental defense of the horse by a few cavalrymen in the 1930s may seem to prove that the cavalry was conservative and backward, yet that Colonel Blimp image, while widespread, did not reflect the reality: a military branch trying to adapt to new technologies by adopting them. Although some cavalrymen were reactionary and backward in their defense of the horse from mechanization, the more common response to the challenges of technology in the 1930s was the continuation of the attitude of acceptance that had marked the cavalry's approach to aviation in the 1920s. Doctrinal statements, student papers, school curricula, cavalry journal articles, and miscellaneous reports show both American and British cavalrymen working with aviation as much as they were able. Circumstances made the Britain cavalry less able to cooperate with air units than their American counterparts, who fought for aviation into the 1940s despite obstruction from air force personnel and politicians supporting independent aviation.

Cavalrymen used similar arguments against mechanization as they had against aviation, but their resistance to these innovations was not an indication of antitechnological or reactionary thinking, but simply a prudent caution in the face of unproven novelty. Among their warnings about exaggerated predictions and technological limitations, cavalrymen argued that mechanization could strengthen the cavalry and be just as or even more helpful than airplanes. Yet mechanization proved the more transformative threat. Budget cuts and mechanized vehicles such as tanks, combat cars, and armored cars proved unstoppable enemies to the horse cavalry. By the end of the 1930s, the British had completely converted to mechanization, except for a few ceremonial regiments. The Americans followed that path more slowly, finally retiring their horse cavalry at the end of World War II.

Conclusion

In his dissertation on cavalrymen and mechanization, Alexander Bielakowski, in the chapter titled "Opponents of Mechanization," accused John K. Herr, the last commander of the American horse cavalry, of being "opposed to the new paradigm of mechanization," and also claimed that Herr's support for the retention of the horse was motivated by nothing more than "personal and emotional attachments to the symbol of [his] . . . profession."[1] However, this explanation is simplistic. Herr's predictions about needing the horse in the future have, in a way, come true. After the American cavalry was officially dismounted in the early 1940s, he warned that horses had not outlived their usefulness. Unless horsemanship training restarted, he feared the knowledge needed for campaigns in regions where terrain and other factors made the new modern technology (jeeps, trucks, motorcycles, airplanes, and tanks) unusable would be lost. Herr realized that the skills required for mounted action would be difficult to re-create on demand.[2] He was correct. The United States military has needed horses since the cavalry was disbanded, most recently in Afghanistan, and the loss of the necessary skills has adversely affected performance, as evidenced by the mishaps of an American soldier who repeatedly fell off of his horse, as reported by the PBS television series *Frontline*.[3] In another case, a special forces team mounted on Afghan ponies was only saved from falling off steep cliffs by the knowledge of its commander, who happened to be a high school rodeo rider in his younger years. American special forces in Afghanistan used horses with wooden saddles as well as camels to transport men and materials through and to conduct reconnaissance in terrain too difficult to cross in Humvees and tanks.[4]

The cavalrymen's prediction that horses would be useful in future conflicts should not be automatically dismissed as an unreasonable or illogical attachment to the past or as a sentimental regard for animals. The expense of maintaining a horse cavalry over the past seventy years for only occasional use would probably not have justified the expense required, but cavalrymen were correct in asserting that horses would be useful in future wars. Labeling cavalry officers as sentimental or antitechnological for listing the limitations of new technologies or defending the continued utility of horses is too simplistic. Cavalrymen recognized the value of aviation and mechanization, but they rejected the contentions that these innovations made the horse cavalry unnecessary. Regardless of the tendency of a few cavalrymen to defend the horse sentimentally, the cavalries of the United States and Britain continually examined and attempted to incorporate new technologies throughout the beginning of the twentieth century. Even Herr called for a combination of horse and mechanized cavalry units.

Herr and his fellow cavalrymen understood that the cavalry would still provide value in situations where airplanes and mechanized vehicles could not be employed, as was demonstrated during the Great War and in maneuvers up through the 1930s. Cavalry journals, committee reports, and other publications show an officers' corps not opposed to the airplane, aviation, aviators, or mechanization advocates on principle; instead, those publications show a group attacking the belief that the cavalry was obsolete due to new technological advancements. Discussions of the limitations of new technologies had less to do with a distrust or hatred of technology and more with debunking the theories of overly optimistic supporters of modern war weapons, elicited in part by national movements for economy and modernization. The cavalry desired to incorporate aviation technology well into the 1930s in Britain and until the early 1940s in the United States. The desire was clear in the cavalry's testing of autogiros. The branch was still attempting to gain control of its own aircraft to maintain its usefulness when World War II commenced.

The defense of the horse in the 1930s may seem to prove that the cavalry was conservative and backward, yet that does not explain trials of experimental aircraft during those same years. Although the cavalry may have gained its reputation of backwardness due to its defense of the horse, the 1930s also provided the example of its push for new innovations, especially its experimentation with new types of aircraft. Rather than criticize cavalrymen's lack of vision about the future impact of airplanes and mechanization or their attachment to horses, it is better to understand the cavalry's experience with airplanes during the more than thirty years the two coexisted.

American and British cavalrymen encountered advancing technologies that had the potential to change war, which would have directly altered the cavalry's armament, tactics, and overall role. They did not reject these technologies outright, but they did not embrace them unquestioningly. Their position, that they wanted to subject the innovations to tests in the field before adopting or rejecting them, cannot be faulted. Cavalry officers were cautious. They were not Luddites, who rejected and attempted to destroy the technology that challenged their way of life, but realists who cautiously examined the capabilities and utility of the innovations for their branch.

Before the turn of the twentieth century, American and British cavalrymen were modernizing their branches to keep pace with changing military conditions that included a variety of possibly revolutionary technologies. Before 1903, British and American cavalrymen actively participated in reforms to meet the demands of modern warfare, through professionalizing their branches, founding military service schools and professional societies, studying military history and recent conflicts, and publishing journals intended to be forums for the discussion of their problems and opportunities. The responses of the United States and Great Britain to modern conditions were influenced by their respective histories, traditions, and culture. The differences between these two nations and within each country's mounted forces were based on the age and history of their branches, the organization of their mounted forces, their wartime experiences, public opinion, and their differing approaches to professionalization.

American and British cavalrymen did not entirely share the same understanding of the roles, armaments, and purposes of their branches. The American cavalry was a much younger and more flexible organization. During its short history, it had been utilized mostly as a dragoon force, equally comfortable fighting mounted or dismounted. American cavalrymen also recognized the importance of their screening, scouting, and reconnaissance functions to the cavalry's continued utility to the army. Their pragmatic doctrine was based on current conditions and not historical examples of massed cavalry charges, as was too often the case with their European counterparts. Since its inception, American cavalry had not fit the traditional definition of the cavalry as primarily a mounted battlefield charging force with its leaders expecting its further development to continue along the same line.

In contrast, British cavalrymen were members of a service that had existed for centuries and had built their pride and spirit around their use of the knee-to-knee charge. Although they were trying to reform themselves to address modern war conditions, disagreements abounded. Throughout the early twentieth century, British cavalrymen engaged in a long and passionate debate about whether the branch was to fight as mounted infantry or as cavalry utilizing sword and lance *en masse* in the *arme blanche.* The British cavalry focused on internal disagreement involving issues of training, tactics, and equipment, greatly hindering its ability to fight a united battle against the outside danger of aviation taking over its reconnaissance functions. British cavalrymen also had to defend the uniqueness of its branch— mobility and the use of arme blanche tactics—against other supposedly cheaper mounted units.

American and British cavalrymen from 1903 until the Great War responded to the potential loss of roles and missions to aviation differently due to their divergent assessments of the importance of reconnaissance, the unique threats expressed in newspapers and periodicals, their dissimilar experiences with aviation, and their proximity to potentially hostile neighbors, which affected how each country estimated the urgency of war preparations. The American cavalry was consistently skeptical of aircraft, whereas the British cavalry lacked a consistent unified position and was willing to allow aircraft to take over its reconnaissance duties, which it considered a low priority. Beyond the theoretical debates, individuals in the American and British militaries addressed the real-world challenges of technologies in transition when no one was certain what airplanes could actually accomplish in the present or near future.

The American and British cavalries' responses to the airplane prior to, during, and after the Great War demonstrated that they were rationally cautious about the ability of aviation to perform and take over cavalry duties. Although both cavalries had little experience with the new technology, they saw it as having far too many limitations and drawbacks to fulfill the predictions made by its supporters. Yet they discussed it, experimented with it in maneuvers, and eventually realized it might be a useful auxiliary to the mounted branch. Some cavalrymen even welcomed aviation to replace or supplement what they considered the less prestigious cavalry functions, once airplanes became developed enough to fill these roles.

Newspaper coverage of airplanes in the United States and Great Britain varied. Military and civilian analysis of the technical primitiveness and limitations of early aircraft received the most attention in Britain, whereas across the Atlantic the general attitude was that airplanes would revolutionize warfare. Predictions that airplanes would soon replace the cavalry quickly appeared in the American press. The lack of direct evidence for aviation's capabilities in no way impeded the enthusiasm of writers who attacked those who did not immediately and wholeheartedly embrace the new technology. More temperate news articles that discussed airplanes' abilities to assist and cooperate with the cavalry and other military branches did not have the impact of the more sensational accounts.

After initially ignoring the airplane, American cavalrymen gave greater attention to airplanes following a 1912 movement to reduce their branch. They realized it was a new tool to be considered seriously in any debate about roles, strategy, tactics, and planning. Their responses included the cautiously optimistic belief that airplanes would make a valuable and possibly essential adjunct to cavalry in reconnaissance and scouting. However, flying machines still suffered from the same shortcomings that had been identified in the years before 1912, such as their inability to function in poor weather and bad terrain.

As aircraft became more reliable and capable each year, American cavalrymen began to admit the possibility of conceding their reconnaissance role to aviation altogether. However, like those before them in the *Journal of the United States Cavalry Association,* they attempted to turn this loss into a victory. Cavalrymen argued that losing this duty could strengthen the cavalry by freeing it up for other important missions and denied that relinquishing this role made the cavalry obsolete or diminished its importance. Reports from military operations and maneuvers confirmed American cavalrymen's beliefs that airplanes could not entirely replace the cavalry in reconnaissance roles.

Unlike their American colleagues, British cavalrymen responded quickly to the airplane in print. The proximity to Britain of other nations developing aviation, especially France and Germany, made the new technology hard to ignore. The British cavalry's treatment of airplanes before 1912 was largely positive, arguing that it would benefit the cavalry if airplanes assisted in or even appropriated its reconnaissance role. Yet, despite seeming willing to share if not cede its reconnaissance roles before 1912, British cavalrymen actively fought the idea that aviation would make them obsolete. Unlike the Americans, some British cavalrymen insisted that reconnaissance was not their branch's primary role.

Following the Great War charges that the cavalry had either functionally or financially outlived its usefulness compelled cavalrymen to respond.

Cavalrymen used examples from the war to support their current usefulness. They also participated in maneuvers to demonstrate their value on a battlefield composed of a combination of forces and challenged the ability of aviation to take over its reconnaissance roles completely.

By late 1919, American cavalrymen were working with the Army Air Service border patrol to solve the problem of communication between the air and ground units revealed during their joint border patrols. Although short-lived, the patrols provided an opportunity for the development of cooperation between air and ground forces in the continental United States. Reports of several maneuvers in the 1920s portray an American cavalry seeking to improve understanding and cooperation between the cavalry and air corps. The Army Air Service partially reciprocated cavalry attempts at cooperation.

British doctrine also supported the importance of combined forces, yet conflict emerged as a result of the organizational structure of the British forces. The American situation was an internal army matter, whereas the British system involved two separate services, the army and air force. The arrangement made combined training in the postwar military more difficult to coordinate because of the additional administrative steps involved. However, British combined training was more structured than its American counterpart. Early maneuvers provided concrete information on the capabilities and limitations of the cavalry and the air services in cooperation and demonstrated the need for the cavalry to fill the holes in aviation employment. Some cavalrymen desired an even greater integration of aviation into existing branches. Combined training led some cavalry officers to think that simply attaching an air unit to a cavalry division was insufficient.

The maneuvers demonstrated that additional cooperation was needed between all units, including the cavalry and aviation. These operations supported the conclusion that neither the cavalry nor aviation could fulfill the roles of the other. Early conclusions from the lessons of the war and combined training led to the creation of manuals, field service regulations, and military school handbooks that provided detailed explanations of the joint uses of aviation and the cavalry in communication, reconnaissance, and security. The core lesson was the importance of combined operations among all military units.

As time passed, however, the cavalry's position lost support in military policy circles, especially in Great Britain. Cooperation between air and ground forces in Great Britain was problematic. Despite the cavalry's willingness to work with the Royal Air Force (RAF), aviators were no longer interested in discussing or training for joint operations. The head of the RAF played down its cooperative roles with the cavalry, claiming that if reductions in its size were necessary, they should be applied to units assisting the army and the cavalry.

Despite American and British cavalrymen's efforts to modernize and cooperate closely with the developing aviation services, unceasing postwar attacks against the cavalry forced some cavalry officers to focus their attention on defending the continued need for their branch. These officers resurrected the old prewar arguments to defend the continued necessity for the cavalry, such as blaming the tendency of Americans to accept new technologies over old ones too readily and enumerating the limitations of airplanes in certain weather and terrain conditions. In addition, as before the war, cavalrymen advanced the belief that aviation was not an enemy of the cavalry, but a technology that strengthened mounted units.

While many cavalry arguments were inspired by the technological limitations of aircraft at the time, one emotional argument existed. This resurfacing argument, popular with cavalrymen of all ranks, was the faith in the man or the man and his horse over the machine, the contrast of the moral superiority of the cavalryman and his horse to the cold, unfeeling flying machine. The belief in the soldiers' fighting power was not confined to a few or limited to the cavalry. This spirit of man overcoming all technological innovations rarely appeared in prewar military writings, it but blossomed during the 1920s.

During that decade, cavalrymen continued to work on modernizing their organization and testing new technologies. They were forced to respond to frequent accusations that the use of airplanes, tanks, and other modern weapons in the recent war proved that the horse cavalry was an obsolete arm with no future. Cavalrymen rejected the idea that these new advancements made the cavalry outdated. Instead, both the American and British cavalries contended that while aviation could accomplish some of the cavalry's roles, such as strategic reconnaissance, many situations still required cavalry, as had been demonstrated during the Great War and in maneuvers and exercises during the 1920s. Cavalry journals, committee reports, and other publications as well as training, maneuvers, and doctrine show a cavalry not rigidly opposed to aviation but a group actively figuring out how to work with these new technologies. Discussions of the limitations of new technology had less to do with a distrust or hatred of technology and more with debunking the theories of overly optimistic supporters of aviation, elicited in part by movements for economy and modernization.

The greatest divergence between the American and British cavalries' experience with aviation occurred in response to their national commitments. Central in the debates over economy and modernization in Britain were attempts to police the British Empire and mandated territories in the 1920s and 1930s. A concerted campaign supported aviation at the expense of the horse cavalry, claiming the former was more effective and cheaper. The Americans had no such requirement.

Although it was becoming clearer that recent innovations would eventually replace the cavalry, cavalrymen maintained that the day still had not arrived, and they were supported by senior military personal and government officials who testified in front of various committees in control of military expenditure that the cavalry was still useful and that further reductions would damage it. The committee members, however, appeared to already have their minds made up and called for a reduction of the cavalry. As reductions commenced, they were stalled on a few occasions when officers warned of the danger of overhasty reductions. Unfortunately for the witnesses and their supporters, their testimony did little to halt the cavalry's continued reduction. In the end, air advocates won the public relations and budget battles. Their assertions that air policing was cheaper than ground and mounted troops were accepted. Although it became clear upon deeper analysis that the success of independent air policing was a myth, the contemporary belief in its value provided support to maintain the RAF and to continue to advocate for reducing national expenditure by reducing ground forces instead of air forces.

Unlike in Great Britain, the need to patrol the U.S.-Mexico border elicited demands for increased cavalry establishments for a few years after the end of the Great War. U.S. attempts at postwar economy hindered aviation's development. The connection between aviation and the cavalry was not entirely ignored in the United States, but a direct connection was often neglected.

By the late 1920s, aviation ceased to be seen as the cavalry-killing technology. Like other innovations such as machine guns, bicycles, and motorcycles, aviation had altered the cavalry's roles and employment, but the mounted branch continued to exist and perform most of its traditional missions. Instead of replacing the cavalry, these innovations worked cooperatively with the branch in reconnaissance, security, communication, protection, and pursuit, an arrangement tested in maneuvers and officially blessed in both British and American doctrine. However, just as cavalrymen were becoming dependent on airplanes, this relationship was altered drastically. The change centered on the shift in aerial doctrine to focus on independent strategic bombardment instead of the tactical support of ground troops. Strategic bombing took the preeminent position in the doctrine and theory of both the Air Corps in the United States and the RAF in Britain, which changed the tone and content of the cavalry debates in the 1930s. These air arms no longer sponsored the development of aircraft for tactical support, nor did they apply much effort to the creation of doctrine for tactical aviation.

As a result, the cavalry demonstrated its progressiveness by attempting to find a replacement for the roles appropriated and then neglected by the air force. Tactical aviation, particularly observation and liaison duties, required slower-moving aircraft than those under development, as slower landing and operational speeds allowed for closer support of troops. The autogiro became the focus of ground forces' attempts to develop their own tactical air assets. Despite its possibilities, the autogiro was a technological failure because it was never widely adopted commercially or militarily. Nevertheless, the cavalry's intensive testing of the experimental aircraft demonstrated a desire to develop and deploy an aerial tool for its missions. Instead of aircraft threatening the cavalry, the American and British cavalries experimented with new air machines to maintain their existence.

By the early 1930s, mechanization in the form of tanks and motor cars replaced aviation as the new technological challenges to the cavalry. Although previous technological innovations had not replaced the cavalry entirely, mechanization advocates circulated familiar predictions about the ability of the new technologies to replace the horse cavalry. Cavalrymen used the example of aviation to show how, yet again, the technological promoters were mistaken.

The cavalry responded to mechanization in much the same way that it replied to aviation. The difference was that mechanization, unlike aviation, directly threatened the horse yet provided an opportunity for the cavalry to retain its roles and élan—but only by transforming itself. As it had done with aircraft in the 1920s, the cavalry tested the possibilities and use of armored cars and tanks in training maneuvers in the 1930s. As they had concluded previously with airplanes, cavalrymen argued that the maneuvers showed the importance of cooperation between horsed and mechanized units, not that the latter replaced the former. The cavalry listed major limitations such as operational difficulties in certain terrain, a dependence on bases of fuel supply, and a need to carry spare parts for field repairs.

Most cavalrymen were excited by the opportunity to evaluate the new technologies' capabilities in maneuvers and operations instead of merely trusting the opinions of salesmen or the public. Although regulations and practice demonstrated that aviation did not make the cavalry obsolete, the contention that the cavalry and its leaders were backward and conservative took on a powerful visual in the form of Colonel Blimp of the British cavalry, an image that lasted for many decades. Wearing his traditional uniform and spurs while riding in his tank, Colonel Blimp became the representation of a coherent reactionary outlook. Characterizing all cavalrymen who defended the continued utility of the cavalry and horse as hidebound anti-intellectuals does a disservice to the majority, who pragmatically addressed new technologies. However, there were some cavalrymen who resorted to sentiment and emotion in the face of the seemingly inexorable pace of progress. These romantic antitechnological arguments may be the cause of the conservative stereotype attributed to many cavalry officers by historians and contemporaries over the past century. But the conservative stereotype fails to accurately describe the vast majority of cavalrymen who enthusiastically tested the new technologies' capabilities.

Unlike in the debates about aviation in the 1920s, the discussion about mechanization and the cavalry in the 1930s centered not on whether the cavalry branch was obsolete but whether the cavalry should be horsed or mechanized. This debate garnered a few extreme responses as it challenged the horse's future. The arguments and terms used in this decade increasingly tied the threat to the cavalry to the defense of the animal over the machine. It is easy to see why in this period cavalrymen were open to attacks of being unthinking, antitechnological, and conservative. Yet focusing on the defense of the traditional horse cavalry masks the true nature of the cavalry's response to technology. The tactic was uncharacteristic of previous arguments made during the first twenty years of military aviation.

There is an irony in that much of the criticism of cavalrymen ("the old") in the early twentieth century was originated by aviation advocates ("the new"), whose twenty-first-century successors are facing a similar challenge of their own. In recent years, the United States Air Force has been experiencing similar debates regarding unmanned aerial vehicles (UAVs). As in the early days of aviation, many politicians and military officers have praised the capabilities of the new technology to improve reconnaissance and conduct precision strikes on enemy targets without putting American lives in jeopardy. However, air force fighter pilots have responded negatively to the effect this technology is having on their position and preparedness of their service for future wars. A September 19, 2009, *Newsweek* article entitled "Attack of the Drones" states that a "fierce fight is on for the mission, culture, and identity of the Air Force, and the Top Guns are losing." Drones, or UAVs, are cheaper than manned aircraft, and their adherents claim they have the same capabilities. Their opponents argue that without a human in them to make on-the-spot, instantaneous decisions, drones are less flexible than aircraft with crews. The article notes that the "Air Force meant fast, agile planes dogfighting high in the sky" to supporters of the F-22 fighter plane. The belief that fighter planes were necessary for future conflicts against other industrialized nations made attempts to eliminate F-22s in favor of drones "tantamount to killing the Air Force." Fighter-pilot generals worry that there will not be enough F-22s in a future war between large air powers.[5] This argument mirrors that of horse cavalrymen who argued that at least one regiment of horse cavalry should be retained in case it was needed in the future.

This is not the first time that the air force has fought a similar battle over its service's missions. General Curtis LeMay, who in the 1950s led the Strategic Air Command, the American bombing force, was opposed to the "development of the intercontinental ballistic missile, which he feared would supplant the long-range bomber. He did not want the Air Force to become 'the silent silo-sitters of the '60s.'"[6] Yet in neither of these cases were air force pilots or bomber pilots called Luddites for their caution with new technologies. The air force has never been characterized as an antitechnological organization, nor have there been calls to put UAVs under the control of a new service branch. Fighter pilots are unlikely to join horse cavalrymen or Luddites as typical examples of backwardness or reactionary stupidity in responding to new technologies in the near future because of the air force's clearly technological culture, but other such cautious technological examiners of a new technology may not be so fortunate. The hope is that this project encourages scholars to examine more groups of technological examiners as well as to avoid missing the journey of technological change and incorporation by concentrating on the conclusion.

Military Aircraft, 1908–1929

	MAX SPEED (mph)	ENDURANCE	CEILING (ft.)	WEIGHT (lbs.)	ENGINE HORSEPOWER	TYPE
Wright 1908 Military Flyer	N/A	N/A	N/A	773	35	General
Farman III, 1909	37	N/A	N/A	1,213	50	General
Curtiss Model D, 1911	50	2 hours, 30 minutes	N/A	1,300	40	General pusher
BE2, 1912	72	3 hours, 15 minutes	10,000	2,350	90	Recon
Farman MF.11 Shorthorn, 1914	66	3 hours, 45 minutes	12,467	2,045	100	General
Curtiss JN-3 (Jenny), 1916	75	2 hours, 15 minutes or 250 miles	N/A	2,130	90	General
Dayton Wright DH4, 1917	124	3 hours, 3 minutes at half throttle	19,000	2,020	230	Recon/light bomber
Packard–Le Pere LUSAC, 1918	136	320 miles	20,200	3,745	425	Fighter
Sopwith 7F.1 (Snipe), 1918	121	3 hours	19,500	2,020	230	Fighter
Handley Page V/1500, 1918	99	17 hours or 1,300 miles	11,000	30,000	375 (4 engines)	Heavy bomber
Vickers Vimy, 1919	100	900 miles	7,000	10,884	200 (2 engines)	Heavy bomber
Martin NBS-1, aka Martin MB-2, 1920	99	400 miles	7,700	12,027	420 (2 engines)	Heavy bomber
Armstrong Whitworth Siskin IIIA, 1925	156	150 miles	27,000	3,012	450	Fighter
Curtiss P-6E Hawk, 1927	198	285 miles	24,700	3,430	600	Fighter
Boulton and Paul Sidestrand, 1928	140	520 miles	24,000	1,060 (bombs)	460 (2 engines)	Medium bomber
Bristol Bulldog II, 1929	178	275 miles	29,300	3,490	490	Fighter

SOURCES: Data from Francis Crosby, *A Handbook of Fighter Aircraft* (London: Hermes House, 2002); Peter Lewis, *The British Bomber since 1914: Fifty Years of Design and Development* (London: Putnam, 1967); Kenneth Munson, *Fighters: Attack and Training Aircraft, 1914–1919* (London: Bounty Books, 2004); Kenneth Munson, *Bombers: Patrol and Reconnaissance Aircraft* (London: Bounty Books, 2012); Justin D. Murphy and Matthew McNiece, *Military Aircraft 1919–1945: An Illustrated History of Their Impact* (Santa Barbara, Calif.: ABC-Clio, 2009); Michael J. H. Taylor, *Jane's Encyclopedia of Aviation* (New York: Portland House, 1989); and Jim Winchester, *American Military Aircraft: A Century of Innovation* (New York: Barnes and Noble Books, 2005).

Reconnaissance Missions

STRATEGIC

MISSION	SERVICE BEST EQUIPPED TO PERFORM MISSION AND MEANS USED	REASON
Areas of concentration of enemy	Air (visual)	These areas are denied to formed troops in the early stages of the combat; other sources are more or less unreliable.
Enemy strength and general composition	Air (visual)	The Air Service, for the same reason, can secure data more nearly correct than can be secured from observers, in most cases, in the early stages of an operation; the information from all sources must be collated.
Routes and direction of movement of each of enemy's main columns	Air (visual and photographic)	Ground troops are often deceived as to the main effort unless they are able to penetrate very definitely the enemy's cavalry screen.
Progress, depth, and width of movement	Cavalry (contact elements)	Can best determine the width and progress of the movements by the establishment and maintenance of contact.
	Air (visual and photographic)	Can best determine the depth of the movement for reasons outlined above.
Location and configuration of enemy's position and his defensive organization	Cavalry (intelligence personnel) Air (visual and photographic, especially the latter)	Each can determine a part of this requirement, and the photographic reconnaissance of the Air Service in this case is of the utmost value, but in the present stage of photography ground troops *must* supplement the interpretative work of the Air Service by ground interpretation.
Location and strength of general reserves or mass of maneuver	Air (visual) Cavalry (as a result of successful penetration of the enemy cavalry screen)	Difficult to secure by any service, but more available to the Air Service.
Lines of supply and administrative establishments	Air (photographic and visual)	Particularly important that these be photographed and made a part of map information for the use of Artillery and Air Service commanders in long-distance bombardment.
Verification and supplementing of information on: topographic and geographic characteristics of terrain	Air (photographic) Cavalry (intelligence personnel) Engineers (topographers)	One of the pre-eminent functions of the Air Service is reconnaissance of this character, which, after interpretation by trained ground troops, is incorporated in map form.
Economic and political characteristics	This is distinctly a function of ground troops and should be made the subject of special reports by the intelligence personnel of these troops.	

TACTICAL

MISSION	SERVICE BEST EQUIPPED TO PERFORM MISSION AND MEANS USED	REASON
Details of the location, distribution, strength, composition, and movements of the enemy	Cavalry, reconnaissance and combat patrols, scouts, observers Air, visual, except in the location of enemy forces, where photographic is valuable	This is a large order and is one of the most important of the tactical reconnaissance missions; each of the reconnaissance services is nearly co-ordinate, one with the other. The Cavalry is considered first because a great portion of this information is incapable of definite determination without combat.
Location of flanks and local reserves of the enemy	Air, visual, and Cavalry	If the Cavalry has properly accomplished its missions, as determined by the results of the preceding strategic reconnaissance, this information is immediately available at all times; if not, visual reconnaissance by the Air Service must make up the deficiency. Both should supplement each other, however, and it is the duty of both.
Local defensive organization of the enemy Local supply arrangements	Air, photographic, and ground troops	The most efficient solution of this mission is by the aerial photography of the organized area, its interpretation by the Air Service observers, and checking and reinterpretation by ground troops in the course of their reconnaissance work, for final compilation in map form preparatory for the attack.
Equipment, training, physical condition, and morale	Ground troops, especially Cavalry	Vigorous and aggressive cavalry reconnaissance, such as that conducted by the famous leaders of the Confederate cavalry during the Civil War, are the only effective solution of this mission during war of movement. During stabilized situations, all troops have their opportunity to perform this class of reconnaissance. It is impossible for the Air Service to assist here except by deduction from the results of other missions.
Detailed examinations of the terrain	Ground troops, based on map information, supplemented by air photography	Aerial photography is excellent for this purpose, in that it gives an excellent detailed study of the area, which when compared with the map will reveal startling facts, but the interpretation of aerial photographs is an art and many things are subject to some doubt, unless all points are definitely verified by ground troops. Here is one of the greatest opportunities for co-operative activity by the Air Service and Cavalry.
Inquiry into local resources and other information	Due to the general nature of this statement, it necessarily includes all services, but when used in this classification of reconnaissance, the Cavalry is the arm which is in the most favorable position to secure such information, from the very nature of its mission.	

SOURCE: Reproduced from Edward Fickett, "A Study of the Relationship between the Cavalry and the Air Service in Reconnaissance," *Cavalry Journal* (US) 32 (October 1923): 415–418.

NOTES

ABBREVIATIONS

AFB	Air Force Base
AIR	Air Ministry and Royal Air Force Records
AFHRA	Air Force Historical Research Agency, Maxwell Air Force Base, Montgomery, Ala.
CAB	Cabinet Office
CGSC	Command and General Staff College
CJ (UK)	*Cavalry Journal*, United Kingdom
CJ (US)	*Cavalry Journal*, United States
FSR	*Field Service Regulations*
GPO	Government Printing Office
HMSO	His/Her Majesty's Stationery Office
JUSCA	*Journal of the United States Cavalry Association*
MHI	U.S. Army Military History Institute, Carlisle, Pa.
NACP	National Archives 2, College Park, Md.
RG	Record Group
TNA	The National Archives (United Kingdom), London
T	Treasury
WO	War Office

INTRODUCTION

1. Kamen quoted in John Heilemann, "Reinventing the Wheel," *Time*, December 2, 2001, http://content.time.com/time/business/article/0,8599,186660,00.html.

2. Unmesh Kher, "The Segway Sage Speaks," *Time*, August 14, 2006, http://content.time.com/time/business/article/0,8599,1226355,00.html.

3. K. E. Bailes, "Technology and Legitimacy: Soviet Aviation and Stalinism in the 1930's," *Technology and Culture* 17 (January 1976): 55–81; Russell I. Fries, "British Response to the American System: The Case of the Small-Arms Industry After 1869," *Technology and Culture* 16 (July 1975): 377–403; Robert C. Post, "The Page Locomotive: Federal Sponsorship of Invention in Mid-19th-Century America," *Technology and Culture* 13 (April 1972): 140–69. Studies of outright technological failures, including the technology's disappearance and loss of investors, are much fewer in number and include the following: Carlos Flick,

"The Movement for Smoke Abatement in 19th-Century England," *Technology and Culture* 21 (January 1980): 29–50; Stuart W. Leslie, "Charles F. Kettering and the Copper-Cooled Engine," *Technology and Culture* 20 (October 1979): 752–76; John H. Perkins, "Reshaping Technology in Wartime: The Effect of Military Goals on Entomological Research and Insect-Control Practices," *Technology and Culture* 19 (April 1978): 169–86; and Michael J. Neufeld, *The Rocket and the Reich: Peenmunde and the Coming of the Ballistic Missile Era* (New York: Free Press, 1995). For more works on technological failures, see the bibliography in John M. Staudenmaier, *Technology's Storytellers: Reweaving the Human Fabric* (Cambridge, Mass.: MIT Press, 1985).

4. David Edgerton, *The Shock of the Old: Technology and Global History since 1900* (Oxford: Oxford University Press, 2007), vi, 210; Jonathan Coopersmith, "Failure & Technology," *Japan Journal for Science, Technology & Society* 18 (2009): 94.

5. One particularly clear example of assuming that a technology was bound to succeed can be found in I. B. Holley Jr., *Ideas and Weapons: Exploitation of the Aerial Weapon by the United States during World War I; A Study in the Relationship of Technological Advance, Military Doctrine, and the Development of Weapons* (New Haven, Conn., Yale University, 1953).

6. Adrian Randall, "Reinterpreting 'Luddism': Resistance to New Technology in the British Industrial Revolution," in *Resistance to New Technology: Nuclear Power, Information Technology, and Biotechnology,* ed. Martin Bauer (Cambridge: Cambridge University Press, 1995), 58.

7. Edward Katzenbach, "The Horse Cavalry in the Twentieth Century: A Study in Policy Response," *Public Policy* 8 (1958): 139.

8. Ibid., 140.

9. Joseph J. Corn, *The Winged Gospel: America's Romance with Aviation, 1900–1950* (New York: Oxford University Press, 1983), 43.

10. Works on the cavalry that discuss a connection between aviation and the cavalry for a few paragraphs or pages include the following: George T. Denison, *A History of Cavalry from the Earliest Times with Lessons for the Future,* 2nd ed. (London: Macmillan, 1913); Louis A. Dimarco, *War Horse: A History of the Military Horse and Rider* (Yardley, Pa.: Westholme, 2008); Peter Newark, *Sabre and Lance: An Illustrated History of Cavalry* (Poole, Dorset, UK: Blandford Press, 1987); Gregory J. W. Urwin, *The United States Cavalry: An Illustrated History* (Pode, UK: Blandford, 1983); Philip Warner, *The British Cavalry* (London: J. M. Dent, 1984); George F. Hofmann, *Through Mobility We Conquer: The Mechanization of U.S. Cavalry* (Lexington: University Press of Kentucky, 2006); Roman Jarymowycz, *Cavalry from Hoof to Track* (Westport, Conn.: Praeger, 2008); David E. Johnson, *Fast Tanks and Heavy Bombers: Innovation in the U.S. Army, 1917–1945* (Ithaca, N.Y.: Cornell University Press, 1998); Harold Winton, *To Change an Army: General Sir John Burnett-Stuart and British Armored Doctrine, 1927–1938* (Lawrence: University Press of Kansas, 1988); Sarah Janelle Rittgers, "From Galloping Hooves to Rumbling Engines: Organizational Responses to Technology in the U.S. Horse Cavalry" (PhD diss., George Washington University, 2003); and Richard Wormser, *The Yellowlegs: The Story of the United States Cavalry* (Garden City, N.Y.: Doubleday, 1966).

11. Airplane replaced aeroplane in common usage in the United States after the term

was adopted by the National Advisory Committee for Aeronautics in 1916; the term airplane was only occasionally utilized in British English. This work uses the term "airplane" when referring to heavier-than-air aircraft except in direct quotation. *Oxford English Dictionary Online,* s.v. "airplane," accessed January 7, 2013, http://www.oed.com/view/Entry/4411#eid8019423.

12. The following are works in aviation history that briefly address the connection between aviation and the cavalry: Dallas R. Brett, *History of British Aviation, 1908–1914,* 2 vols. (London: Aviation Book Club, 1934); Charles de Forest Chandler and Frank Purdy Lahm, *How Our Army Grew Wings: Airmen and Aircraft before 1914* (New York: Ronald Press, 1943); Holley, *Ideas and Weapons*; Johnson, *Fast Tanks*; Peter Mead, *The Eye in the Air: History of Air Observation and Reconnaissance for the Army, 1785–1945* (London: HMSO, 1983); Malcolm Smith, *British Air Strategy between the Wars* (Oxford: Clarendon, 1984); and Lynn Montross, *Cavalry of the Sky: The Story of U.S. Marine Combat Helicopters* (New York: Harper and Brothers, 1954).

13. Robin Higham, *100 Years of Air Power & Aviation* (College Station: Texas A&M University Press, 2003); William O. Odom, *After the Trenches: The Transformation of U.S. Army Doctrine, 1918–1939* (College Station: Texas A&M University Press, 1999); David E. Omissi, *Air Power and Colonial Control: The Royal Air Force, 1919–1939* (Manchester, UK: Manchester University Press, 1990); Elwood L. White, *Air Power and Warfare: A Supplement* (Colorado Springs, Colo.: United States Air Force Academy, 2002); Stephen Budiansky, *Air Power: The Men, Machines, and Ideas That Revolutionized War, from Kitty Hawk to Gulf War II* (New York: Viking, 2004); William Carr Sherman, *Air Warfare* (New York: Ronald, 1926); David R. Mets, *Airpower and Technology: Smart and Unmanned Weapons* (Westport, Conn.: Praeger, 2009); James L. Stokesbury, *A Short History of Airpower* (London: Robert Hale, 1986); James S. Corum and Wray R. Johnson, *Airpower in Small Wars: Fighting Insurgents and Terrorists* (Lawrence: University Press of Kansas, 2003); Thomas H. Greer, *The Development of Air Doctrine in the Army Air Arm, 1917–1941* (Washington, D.C.: Office of Air Force History, 1985); Samuel John Gurney Hoare Templewood, *Empire of the Air: The Advent of the Air Age, 1922–1929* (London: Collins, 1957).

CHAPTER 1

1. Walter Millis, *Arms and Men: A Study in American Military History* (1956; repr., New Brunswick, N.J.: Rutgers University Press, 1984), 21; Russell F. Weigley, *History of the United States Army* (New York: Macmillan, 1967), 70; Urwin, *United States Cavalry,* 17, 29.

2. "Report of the Secretary of War," November 25, 1832, Appendix to the Register of Debates in Congress, *Cong. Globe,* 22nd Cong., 2nd Sess. 8–9 (1832).

3. Millis, *Arms and Men,* 95–96; Mary Lee Stubbs and Stanley Russell Connor, *Armor-Cavalry Part I: Regular Army and Army Reserve* (Washington, D.C.: Office of the Chief of Military History, United States Army, 1969), 8; Weigley, *History of the United States Army,* 159; Urwin, *United States Cavalry,* 54.

4. Stubbs and Connor, *Armor-Cavalry Part I,* 10–11.

5. Alexander M. Bielakowski, *U.S. Cavalryman, 1891–1920* (Oxford: Osprey, 2004), 5.

6. Richard Wormser, *The Yellowlegs: The Story of the United States Cavalry* (Garden City, N.Y.: Doubleday, 1966), 212.

7. Weigley, *History of the United States Army,* 239. See also John Bigelow Jr., *The Principles of Strategy: Illustrated Mainly from American Campaigns,* 2nd ed. (Philadelphia: J. B. Lippincott, 1894), 92–102, 131–32.

8. Urwin, *United States Cavalry,* 133.

9. Stubbs and Connor, *Armor-Cavalry Part I,* 21.

10. Ronald Barr, *The Progressive Army: U.S. Army Command and Administration, 1870–1914* (New York: St. Martin's Press, 1998), 2.

11. Stubbs and Connor, *Armor-Cavalry Part I,* 22.

12. Timothy K. Nenninger, *The Leavenworth Schools and the Old Army: Education, Professionalism, and the Officer Corps of the United States Army, 1881–1918* (Westport, Conn.: Greenwood, 1978), 3–4.

13. Weigley, *History of the United States Army,* 273.

14. Ibid., 290; Nenninger, *Leavenworth Schools and the Old Army,* 34. The school's name and character changed in 1908, becoming the Mounted Service School. In 1919 it was renamed the Cavalry School. (Elizabeth T. Kirwan, "The Cavalry School Library, Fort Riley, Kansas," *Library Journal* 62 [March 1937]: 196.)

15. Carol Reardon, *Soldiers and Scholars: The U.S. Army and the Uses of Military History, 1865–1920* (Lawrence: University Press of Kansas, 1990), 16.

16. *Regulations and Programm of Instruction of the U.S. Infantry and Cavalry School* (Fort Leavenworth, Kans.: United States Infantry and Cavalry School, 1895), 33.

17. Ibid.

18. Weigley, *History of the United States Army,* 273; Reardon, *Soldiers and Scholars,* 14–15.

19. Eben Swift, "An American Pioneer in the Cause of Education," *Journal of the Military Service Institution of the United States* 44 (1909): 67, quoted in Reardon, *Soldiers and Scholars,* 38; T. R. Brereton, *Educating the U.S. Army: Arthur Wagner and Reform, 1875–1905* (Lincoln: University of Nebraska Press, 2000), xii; Weigley, *History of the United States Army,* 274.

20. Arthur L. Wagner, *Organization and Tactics* (New York: B. Westermann, 1895).

21. The following were additional writings used in the military postgraduate schools: William Harding Carter, *Horses, Saddles and Bridles* (Leavenworth, Kans.: Ketcheson and Reeves, 1895); Charles Blair Mayne, *The Infantry Weapon and Its Use* (London: Smith, Elder, 1903); Colmar Freiherr von der Goltz, *The Conduct of War: A Short Treatise on Its Most Important Branches and Guiding Rules,* trans. G. F. Leverson (London: K. Paul, Trench, Trübner and Co, 1899); *Field Service Regulations;* Otto Griepenkerl, *Letters on Applied Tactics* (Kansas City, Kans.: Hudson Press, 1906); *Notes on Infantry, Cavalry, and Field Artillery Lectures Delivered to Class of Provisional Second Lieutenants Fort Leavenworth, Kansas, 1917* (Washington, D.C.: GPO, 1917). Drill regulations of the three arms were also utilized at these schools, including *Drill Regulations for Cavalry, United States Army: Amended 1909, Corrected to January 1* (Washington, D.C.: GPO, 1911); and War Department: Office of the Chief of Staff, *Cavalry Drill Regulations United States Army 1916, Corrected to December 31, 1917 (Changes nos. 1 and 2)* (Washington, D.C.: GPO, 1918).

22. Brereton, *Educating the U.S. Army,* 51–52.

23. Nenninger, *Leavenworth Schools and the Old Army,* 42–43; Reardon, *Soldiers and Scholars,* 37.

24. Arthur L. Wagner, *The Service of Security and Information* (Kansas City, Mo.: Hudson-Kimberly, 1903), 18–19, 100; Arthur L. Wagner, *A Catechism of Outpost Duty, Including Advance Guards, Rear Guards, and Reconnaissance* (Kansas City, Mo.: Hudson-Kimberly, 1895), 126.

25. Wagner, *Catechism of Outpost Duty*, 23, 26.

26. Wagner, *Service of Security and Information*, 31.

27. Ibid., 31–32.

28. Wagner, *Catechism of Outpost Duty*, 113.

29. Wagner, *Service of Security and Information*, 152.

30. Allan R. Millett and Peter Maslowski, *For the Common Defense: A Military History of the United States of America* (New York: Free Press, 1984), 304, 327–28; Stubbs and Connor, *Armor-Cavalry Part I*, 27–30.

31. Nenninger, *Leavenworth Schools and the Old Army*, 103.

32. Wagner, *Service of Security and Information*, 129–133; Army and Navy Journal, "Is the Cavalry Obsolete?" *Hartford Courant*, November 2, 1903, 13.

33. For debates on the saber versus the carbine, see John Bigelow Jr., "The Sabre and Bayonet Question," *Journal of the Military Service Institute of the United States* 3 (1882): 65–96, E. P. Andrus, "The Saber," *JUSCA* 4 (December 1892): 373–82; M. C. Butler, "The Saber," *JUSCA* 14 (July 1904): 142–44; James Parker, "The Retention of the Saber as a Cavalry Weapon," *JUSCA* 14 (October 1904): 354–65; George Vidmer, "The Service Pistol and Its Caliber," *JUSCA* 16 (October 1905): 181–88; and James Parker, "Saber Versus Revolver Versus Carbine," *JUSCA* 17 (July 1906): 35–41. For debates on mounted versus dismounted action, see Charles D. Rhodes, "The Duties of the Cavalry in Modern Wars," *JUSCA* 6 (June 1893): 172–81; George H. Morgan, "Mounted Rifles," *JUSCA* 13 (January 1903): 380–85; and J. A. Augur et al., "Comments on Mounted Rifles," *JUSCA* 13 (January 1903): 386–407. For additional debates on firearms, see Aubrey Lippincott, "The Automatic Small Arm," *JUSCA* 13 (July 1902): 66–70; and Thomas Q. Donaldson, "The Revolver or Pistol Best Suited to Cavalry," *JUSCA* 13 (July 1902): 71–74. For the value of shock action, see J. Y. Blunt, "The Shock Action of Cavalry," *JUSCA* 5 (1892): 33–45; J. A. Augur, Edward Anderson, and Cornelius C. Smith, "Comment and Criticism-Mounted Rifles," *JUSCA* 13 (April 1903): 718–23; and James G. Harbord, "Cavalry in Modern War," *JUSCA* 15 (April 1905): 765–71.

34. Barr, *Progressive Army*, 2.

35. Moses Harris, "With the Reserve Brigade—From Winchester to Appomattox. Fourth and Concluding Paper." *JUSCA* 4 (March 1891): 25.

36. Urwin, *United States Cavalry*, 108.

37. Ibid.

38. G. F. R. Henderson, *The Science of War: A Collection of Essays and Lectures, 1891–1903* (New York: Longsmans, Green, and Co., 1908), 51.

39. Stephen Badsey, *Doctrine and Reform in the British Cavalry, 1880–1918* (Aldershot, England: Ashgate, 2008), 4–6, 10.

40. Ibid., 5.

41. Brian Bond, "Doctrine and Training in the British Cavalry, 1870–1914," in *The Theory and Practice of War*, ed. Michael Howard (London: Cassell, 1965), 103.

42. Badsey, *Doctrine and Reform in the British Cavalry,* 10.

43. Ibid., 12–13.

44. Ibid., 13–14; Bond, "Doctrine and Training in the British Cavalry," 104.

45. Badsey, *Doctrine and Reform in the British Cavalry,* 15.

46. Ibid.

47. Marquess of Anglesey, *A History of British Cavalry,* vol. 4, *1899–1913* (London: Secker and Warburg, 1986), 408.

48. Badsey, *Doctrine and Reform in the British Cavalry,* 15.

49. Bond, "Doctrine and Training in the British Cavalry," 104–5.

50. Ibid. See also Brian Holden Reid, "'A Signpost That Was Missed'?: Reconsidering British Lessons from the American Civil War," *Journal of Military History* 70 (April 2006): 397.

51. Reid, "A Signpost That Was Missed," 412.

52. Anglesey, *History of British Cavalry,* 163, 232.

53. Ibid., 25; Bond, "Doctrine and Training in the British Cavalry," 108.

54. Bond, "Doctrine and Training in the British Cavalry," 110.

55. Badsey, *Doctrine and Reform in the British Cavalry,* 2.

56. Anglesey, *History of British Cavalry,* 382–83.

57. Ibid., 376, 386.

58. Ibid., 428.

59. Ibid., 25–26. See also Bond, "Doctrine and Training in the British Cavalry," 99.

60. Bond, "Doctrine and Training in the British Cavalry," 99.

61. Badsey, *Doctrine and Reform in the British Cavalry,* 3.

62. Ibid.

63. Books include John Formby, *Cavalry in Action in the Wars of the Future* (London: H. Rees, 1905); Friedrich von Bernhardi, *Cavalry in Future Wars* (London: John Murray, 1909); Friedrich von Bernhardi, *Cavalry in War and Peace* (London: H. Rees, 1910); Erskine Childers, *War and the Arme Blanche* (London: E. Arnold, 1910); and Erskine Childers, *German Influence on British Cavalry* (London: E. Arnold, 1911).

64. Badsey, *Doctrine and Reform in the British Cavalry,* 2.

65. Ibid.

66. Anglesey, *History of British Cavalry,* 256, 394, 413.

67. Badsey, *Doctrine and Reform in the British Cavalry,* 2.

68. Bond, "Doctrine and Training in the British Cavalry," 108.

69. Frederic Natusch Maude, *Cavalry: Its Past and Future* (London: William Clowes & Sons, Limited, 1903), viii.

70. Anglesey, *History of British Cavalry,* 26.

71. Bond, "Doctrine and Training in the British Cavalry," 111–12.

72. Childers, *War and Arme Blanche,* v-xvi; Bond, "Doctrine and Training in the British Cavalry," 116.

73. Henderson, *The Science of War,* 77.

74. Douglas Haig, *Cavalry Studies: Strategical and Tactical* (London: H. Rees, 1907), 3, quoted in Anglesey, *History of British Cavalry,* 431.

75. Notrefe [Douglas Haig], *Cavalry Taught by Experience: A Forecast of Cavalry under*

Modern Conditions (London: H. Rees 1910), 40–42, quoted in Anglesey, *History of British Cavalry*, 431.

76. [Official] *Report on Combined Manoeuvres*, September 1903, 4, quoted in Anglesey, *History of British Cavalry*, 437.

77. Anglesey, *History of British Cavalry*, 444.

78. Herbert A. Johnson, *Wingless Eagle: U.S. Army Aviation through World War I* (Chapel Hill: University of North Carolina Press, 2001), 12; Harold Hinton, *Air Victory: The Men and the Machines* (New York: Harper and Brothers, 1948), 5; John H. Morrow Jr., *The Great War in the Air: Military Aviation from 1909 to 1921* (Washington, D.C.: Smithsonian Institution Press, 1993), 4.

79. Appendix No. 1 Army Officers Ordered to Aeronautical Duty, Chronologically Arranged, n.d., Call No. 167.401-10, 1909–1939, IRIS No. 00120709, Frank Purdy Lahm Collection, 1913–1939, AFHRA; and Frank Lahm, No Title—Summary of Military Aeronautics Plan, February 3, 1909, Call No. 167.601-30, IRIS No. 1010118, in Correspondence File, General Lahm, 1907–1943, AFHRA.

80. Morrow, *Great War in the Air*, 4.

81. United States Air Corps Tactical School, *Observation Aviation March 1930* (Langley Field, Va.: Air Corps Tactical School, 1930), 1; Holley, *Ideas and Weapons*, 26-27.

82. Benjamin D. Foulois with C. V. Glines, *From the Wright Brothers to the Astronauts: The Memoirs of Major Benjamin D. Foulois* (New York: McGraw-Hill, 1968), 2.

83. Ibid., 4–5.

84. Major B. D. Foulois to Lieut. General Robert L. Bullard, Personal Service Record for Period 1898–1919, October 14, 1919, Call No. 168.68–5, 1898–1919, IRIS No. 00125307, in B. D. Foulois Papers, AFHRA.

85. Morrow, *Great War in the Air*, 49.

86. U.S. War Department, *FSR, United States Army, 1914* (Washington, D.C.: GPO, 1914), 13.

87. Morrow, *Great War in the Air*, 50.

88. Air Corps Tactical School, *Observation Aviation*, 1.

89. Joseph J. Corn, ed., *Imagining Tomorrow: History, Technology, and the American Future* (Cambridge, Mass.: MIT Press, 1986), 2.

90. Corn, *Winged Gospel*, 135. For additional discussion of the religious treatment of aviation by Americans, see Robert Wohl, *A Passion for Wings: Aviation and the Western Imagination, 1908–1918* (New Haven, Conn.: Yale University Press, 1994); and Robert Wohl, *The Spectacle of Flight: Aviation and the Western Imagination, 1920–1950* (New Haven, Conn.: Yale University Press, 2005).

91. Corn, *Winged Gospel*, vii.

92. As quoted in Alfred Gollin, *No Longer an Island: Britain and the Wright Brothers, 1902–1909* (Stanford, Calif.: Stanford University Press, 1984), 2.

93. Alfred Gollin, *The Impact of Air Power on the British People and Their Government, 1909-14* (Stanford, Calif.: Stanford University Press, 1989), 49–63. See also Alfred M. Gollin, "England Is No Longer an Island: The Phantom Airship Scare of 1909," *Albion: A Quarterly Journal Concerned with British Studies* 13, no. 1 (Spring 1981): 43–57; and W. Michael Ryan, "The Invasion Controversy of 1906–1908: Lieutenant-Colonel Charles a

Court Repington and British Perceptions of the German Menace," *Military Affairs* 44, no. 1 (February 1980): 8–12.

94. Corn, *Imagining Tomorrow*, 222.

95. Charles C. Turner, "Aeronautics," *Observer,* September 17, 1911, 11. Turner joined with the pioneering British aviator Gustav Hamel to write *Flying: Practical Experience,* expounding on their early experiences with aviation. Unfortunately, Hamel disappeared while flying a new monoplane over the English Channel. For more details, see "Aviator Hamel Lost?: Starts on 100-Mile Trip Across the British Channel—Not Heard From," *New York Times,* May 24, 1914, 11; "Channel Searched No Trace of Hamel," *New York Times,* May 25, 1914, 1; "Body Surely Hamel's: Corpse Found and Abandoned by Fisherman That of Airman," *New York Times,* July 9, 1914, 4.

96. Gollin, *Impact of Air Power*, 26.

97. David Edgerton, *England and the Aeroplane: An Essay on a Militant and Techno-logical Nation* (Basingstoke: Macmillan in Association with the Centre for the History of Science, Technology and Medicine, University of Manchester, 1991), 6.

98. Gollin, *Impact of Air Power*, 4.

99. *Naval and Military Aeronautics*, Hansard HC Deb 02, August 1909, vol. 8, col. 1574.

100. Gollin, *Impact of Air Power*, 12.

101. Hugh Driver, *The Birth of Military Aviation: Britain, 1903–1914* (Woodbridge, UK: The Royal Historical Society: Boydell Press, 1997), 26–27; Ben Mackworth-Praed, ed., *Aviation: The Pioneer Years* (Secaucus, N.J.: Chartwell Books, 1990), 137.

102. Alfred Gollin, "The Mystery of Lord Haldane and Early British Military Aviation," *Albion: A Quarterly Journal Concerned with British Studies* 11, no. 1 (Spring 1979): 60–61.

103. Edgerton, *England and the Aeroplane*, 4.

104. Morrow, *Great War in the Air,* 21.

105. Ibid.

106. "House of Commons," *Manchester Guardian,* August 3, 1909, 6. For more on the airplane's value compared to the dirigible or airship in Britain, see "The New Instrument of War," *Living Age,* November 12, 1910, 441.

107. H. Bannerman-Phillips, "The Future of Airships in War," *United Service Maga-zine* 37 (September 1908): 589.

108. "House of Commons," 6.

109. Dimitar Nedialkov, *Genesis of Airpower* (Sofia: Pensoft, 2004), 178; Eric Lawson and Jane Lawson, *The First Air Campaign: August 1914–November 1919* (Conshohocken, Pa.: Combined Books, 1996), 22; Michael Taylor, *Jane's Fighting Aircraft of World War I* (1919; repr., London: Random House, 2001), 28.

110. Air Ministry-Air Historical Branch (Great Britain), *A Short History of the Royal Air Force* ([S.I.] Air Ministry, 1936), 9.

111. Ibid., 10.

112. Ibid., 11.

113. "Aeronautical Reports for 1910," in *The Development of Aircraft*, n.d., AIR 2/2, TNA.

114. Ibid.

115. Gollin, *Impact of Air Power,* 307.

116. Air Corps Tactical School, *Observation Aviation,* 2.

117. "Aeronautical Reports for 1910."

118. Air Ministry, *Short History,* 15.

119. Gollin, *Impact of Air Power,* 307. All of the numbers of planes are approximations because sources vary with regard to the exact number of functional aircraft in each country. For additional estimates, see Lee Kennett, *The First Air War, 1914-1918* (New York: Free Press, 1991), 21; Lawson and Lawson, *First Air Campaign,* 35–37; Morrow, *Great War in the Air,* 47–57; Nedialkov, *Genesis of Airpower,* 174–75, 182–84; and Edgerton, *England and the Aeroplane,* 10.

120. Edgerton, *England and the Aeroplane,* 10.

121. Ibid., xiv, 7–8.

CHAPTER 2

1. Jarymowycz, *Cavalry from Hoof to Track,* 131.

2. In the early twentieth century, numerous authors mentioned utilizing airplanes for reconnaissance. A small sample includes "House of Commons," 6; "Wright Wins Berlin Away from Zeppelin," *New York Times,* September 12, 1909, C2; "Aeroplanes the Things," *Detroit Free Press,* March 17, 1907, B11; "Predicts War in the Clouds," *Los Angeles Times,* August 4, 1907, I4; and E. ff. W. Lascelles, "The Airship and Flying Machine in War: Their Probable Influence on the Role of Cavalry," *CJ* (UK) 5 (April 1910): 211.

3. Morrow, *Great War in the Air,* 1–3.

4. "Aeroplanes the Things," B11.

5. John P. Wisser, "The Tactical and Strategical Use of Balloons and Aeroplanes," *JUSCA* 21 (November 1910): 413–15.

6. "Foresees Aerial War," *Washington Post,* January 1, 1908, 4; "Friction Over Lahm," *Washington Post,* November 6, 1909, 4; Frank P. Lahm, "The Relative Merits of the Dirigible Balloon and Aeroplane in Warfare," *Journal of the Military Service Institution of the United States* 48 (1911): 202.

7. Walter E. Prosser, "A Discussion of the War Balloon and Similar Craft, and the Best Methods of Attack by Artillery," *Journal of the United States Artillery* 34 (July–August 1910): 258.

8. W. A. Tilney, "Aerial Reconnaissance in War," *CJ* (UK) 6 (January 1911): 13.

9. "Aeroplanes at the Army Manoeuvres," *Times* (London), September 3, 1912, 2. For additional British articles mentioning aerial reconnaissance, see Lascelles, "Airship and Flying Machine," 209; "Editorial Article 2-No Title," *Scotsman,* May 13, 1911, 8; "The Military Aeroplane," *Times* (London), December 7, 1911, 7; "The Military Aeroplane: Attitude of the War Office," *Times* (London), December 19, 1911, 10; and "The Army Aeroplane Accident," *Times* (London), September 9, 1912, 8. For additional American articles discussing aerial reconnaissance, see "Aeroplanes the Things," B11; "Future Wars Will Be Fought in the Air," *New York Times,* October 29, 1907, 10; "Aeroplane in War: French Expert Says It May Supercede Cavalry," *New York Times,* August 14, 1908, 8; "Mishap to Aeroplane," *Washington Post,* August 14, 1908, 2; "The Future of Flying," *Living Age,* September 10, 1910, 694; "The Aeroplane in War," *JUSCA* 21 (November 1910): 536–37; "New Instrument," 267; R. A. Campbell, "Aeroplanes with Cavalry," *JUSCA* 22 (September 1911): 311; Turner,

"Aeronautics," 11; and F. S. Foltz, "The Necessity for Well Organized Cavalry," *JUSCA* 23 (March 1913): 726.

10. "The Future of Cavalry," *Times* (London), April 25, 1912, 10. See also H. de B. de Lisle, "The Strategical Action of Cavalry," *CJ* (UK) 7 (January–October 1912): 320–34 for his lecture given at the Royal United Service Institution [hereafter de Lisle, "Strategical Action," *CJ* (UK)] that can also be found as a reprint from the June 1912 *Journal of the Royal United Service Institution* as H. de B. de Lisle, "The Strategical Action of Cavalry," *JUSCA* 23 (July 1912): 134 [hereafter de Lisle, "Strategical Action," *JUSCA*].

11. "Future of Cavalry," 10. See de Lisle, "Strategical Action," *CJ* (UK), 328; Vindex, "Modern Inventions and the Functions of Cavalry," *CJ* (UK) 7 (January–October 1912): 348; and Edmund Candler, "The Aeroplane in Mesopotamia," *Times* (London), November 2, 1916, 5 for more on airplanes as auxiliary to ground forces.

12. F. H. Sykes, "Report upon the Employment of the Royal Flying Corps in Army Manoeuvres, 1912," November 16, 1912, WO 33/620, TNA.

13. "The Cavalry Division Reconnaissance," *Times* (London), September 9, 1912, 8.

14. "Aeroplanes the Things," B11.

15. "Predicts War in the Clouds," I4.

16. "New Instrument," 267. See similar argument in "Future Wars," 10.

17. "Aeroplane in War," *New York Times*, 8; "Future of Flying," 694. For similar articles, see "Mishap to Aeroplane," 2; "Wright Wins Berlin," C2; Henry Woodhouse, "The Airscout," *Town and Country*, November 25, 1911, 34; "Predicts War in the Clouds," I4.

18. "Wright Wins Berlin," C2.

19. "Sharpshooters in Aeroplanes," *Los Angeles Times*, August 24, 1910, 17.

20. "Editorial," *American Aeronaut* 1 no. 3 (October 1909): 115.

21. "Aeroplanes in Wars of Future: Aviators Declare Air Vessels Must Constitute Cavalry of Army," *Indianapolis Star*, December 24, 1911, C27. For more on discussions of the tension between aviation and the cavalry, see E. H. Gilpin, "Armament and Equipment of the Cavalryman," *JUSCA* 22 (July 1911): 82; "The Aeroplanes and the Cavalry," *JUSCA* 23 (July 1912): 123–24; Nickolaus Riedl, "Cavalry in War," *JUSCA* 23 (September 1912): 290; and N. F. McClure, "The Use of Cavalry," *JUSCA* 24 (May 1914): 965.

22. "Future Wars," 10.

23. Frederick Palmer, "Military Lessons Taught by the War," *New York Times*, December 5, 1912, 2.

24. See also "Wright Has Quiet Day," *Washington Post*, September 14, 1908, 2; "What Could Uncle Sam Do in the Event of Aerial War?" *Baltimore Sun*, February 23, 1913, B1; Henry J. Reilly, "Aerial Service Cannot Replace Cavalry Scouts," *Chicago Daily Tribune*, May 11, 1916, 6; Joseph W. Griggs, "Cavalry and Riders of Air Close Rivals in Staging Thrills," *Nashville Tennessean and the Nashville American*, August 16, 1918, 10; "Allies Win by Putting on Wheel and Wing: See Cavalry of Air and Tanks for All Troops in Near Future," *Chicago Daily Tribune*, August 11, 1918, 3.

25. John R. M. Taylor, "Cavalry and the Aeroplane," *Journal of the United States Infantry Association* 6 (July 1909): 87.

26. Ibid., 84–85.

27. Ibid., 85–88.

28. "Modern War," *JUSCA* 20 (March 1910): 965; Holland Rubottom, "Cavalry Reconnaissance and Transmission of Information by Modern Methods," *JUSCA* 23 (July 1912): 25–26; and Daniel L. Roscoe, "The Effect of Aeroplanes upon Cavalry Tactics," *JUSCA* 24 (March 1914): 857.

29. The following is a sample of articles and letters contained in the *JUSCA* calling for a cavalry chief: Velox [pseud.], "A Chief of Cavalry," *JUSCA* 15 (April 1905): 944–46; Malvern-Hill Barnum, "A Chief of Cavalry," *JUSCA* 15 (April 1905): 946–49; George H. Morgan, "Some Needs of the Cavalry," *JUSCA* 16 (October 1905): 329–31; X [pseud.], "Chief of Cavalry: Shall We Have a Chief of Cavalry?" *JUSCA* 17 (January 1907): 556–58; Howard R. Hickok, "Chief of Cavalry," *JUSCA* 18 (October 1907): 347–50; E. S., "Chief of Cavalry," *JUSCA* 18 (January 1908): 555–56; and Editor's Table, "Chief of Cavalry," *JUSCA* 19 (July 1908): 197–98.

30. "Cavalry and the Aeroplane," *JUSCA* 20 (November 1909): 617.

31. V. Stockhausen, "Airships in War," *JUSCA* 20 (November 1909): 575–84; "The Aeroplane in War," *JUSCA* 21 (November 1910): 536–37; Campbell, "Aeroplanes with Cavalry," 313; Captain Niemann, "Airships and Cavalry in the Reconnaissance Service," *JUSCA* 22 (March 1912): 874–75; M. S. E. Harry Bell, "The First Aeroplane under Rifle Fire," *JUSCA* 22 (May 1912): 1128–30; and Paul Hayne Jr., "Organization and Employment of Cavalry Brigades," *JUSCA* 24 (September 1913): 202.

32. "The Aeroplane in War,"*JUSCA* 21 (November 1910), 536–37. *Broad Arrow* was the nineteenth- and early-twentieth-century periodical for military services in Britain that was later incorporated into the *Army and Navy Gazette*.

33. Ibid., 538.

34. "Modern War," 965. See also Sykes, "Employment of the Royal Flying Corps"; and Barry Domvile, "First Annual Report by the Air Committee on the Progress of the Royal Flying Corps," June 7, 1913, AIR 1/2311/221/6, TNA.

35. Riedl, "Cavalry in War," 290.

36. Boots and Saddles [pseud.], "Cavalry Equipment," *JUSCA* 21 (March 1911): 968.

37. Corn, *Winged Gospel*, 29–42.

38. Gilpin, "Armament and Equipment," 76. See also an earlier warning that cavalry officers should retain their "calmness of mind" and objectively discuss other means introduced to complete their roles. J. F. Reynolds Landis, "Spirit of Sacrifice in Cavalry and Esprit de Corps in Its Officers," *JUSCA* 20 (July 1909): 1215–16.

39. Gilpin, "Armament and Equipment," 82. See also Howard R. Hickok, "Role and Organization of Cavalry," *JUSCA* 25 (July 1914): 75.

40. For detailed coverage of this debate, see Badsey, *Doctrine and Reform in the British Cavalry, 1880–1918*.

41. "War and the 'Arme Blanche,'" *CJ* (UK) 5 (July 1910): 283–87; D. C. Crombie, "Cavalry in Frontier Warfare," *CJ* (UK) 5 (July 1910): 358–69; An Infantry Transfer, "Impressions," *CJ* (UK) 5 (July 1910): 350.

42. "Cavalry and the Mounted Infantry," *JUSCA* 17 (October 1906): 363–66; de Lisle, "Strategical Action," *CJ* (UK), 324, 334; Hickok, "Role and Organization," 73; and John Stuart Barrows, "The Uhlans and Other Cavalry in the European War," *JUSCA* 26 (January 1916): 391, 398.

43. The Inspector of Cavalry, "What Lies Before Us," *CJ* (UK) 1 (January 1906): 3.

44. Ibid., 5–6.

45. Ibid., 10–11.

46. "Preface," *CJ* (UK) 6 (January 1911): B–B2.

47. Ibid.

48. Anglesey, *History of British Cavalry*, 401.

49. M. J. Mayhew and G. Skeffington Smyth, "Motor Cars with the Cavalry Division," *CJ* (UK) 4 (October 1909): 442.

50. H. Bannerman-Phillips, "Air-craft in Co-operation with Cavalry," *Nineteenth Century and After* 69 (1911): 806.

51. Lascelles, "Airship and Flying Machine," 211.

52. Tilney, "Aerial Reconnaissance," 12.

53. Ibid., 13.

54. Bannerman-Phillips, "Air-craft in Co-operation," 806. For more on airplane reliability, see also Lascelles, "Airship and Flying Machine," 211; Sykes, "Employment of the Royal Flying Corps"; and Domvile, "First Annual Report by the Air Committee".

55. Bannerman-Phillips, "Air-craft in Co-operation," 807–10.

56. Lascelles, "Airship and Flying Machine," 208.

57. R. A. Cammell, "Aeroplanes with Cavalry," *CJ* (UK) 6 (April 1911): 197.

58. Ibid.

59. Ibid., 199. See also F. E. Waldron, "Aeroplanes and Cavalry," *CJ* (UK) 8 (January–October 1913): 313; "A Study of Patrol Work," *CJ* (UK) 8 (January–October 1913): 427; and Vindex, "Modern Inventions," 349–50.

60. Cecil Battine, "The German Cavalry Manoevures of 1911," *CJ* (UK) 7 (January 1912): 48–49.

61. Waldron, "Aeroplanes and Cavalry," 314. See also Vedette, "Aeroplanes and Their Influence upon Cavalry Training," *United Service Magazine* 83, no. 1005 (1912): 557.

62. H. Clifton Brown, "The Training of Cavalry in Battle," *CJ* (UK) 7 (January–October 1912): 156.

63. Ibid., 158.

64. Ibid., 156–60.

65. Vindex, "Modern Inventions," 348.

66. Ibid., 349. See also Waldron, "Aeroplanes and Cavalry," 313.

67. Woodhouse, "Airscout," 34; Henry Woodhouse, "The Progress of Aviation," *Independent . . . Devoted to the Consideration of Politics, Social and Economic Tendencies, History, Literature, and the Arts,* June 6, 1912, 1264; "Army Air Scouts Do Great Work," *New York Times,* April 12, 1914, 8; "Aero Aids Maneuver," *Washington Post,* October 16, 1914, 11; "One Aeroplane Equal to Whole Cavalry Division," *Aero and Hydro: America's Aviation Weekly* 9 (October 17, 1914): 26; "Aeroplanes Built to Fly All Over the World," *Wall Street Journal,* January 12, 1916, 8; and Our Aviation Correspondent, "Aircraft in the War," *Observer,* October 5, 1916, 13.

68. Woodhouse, "Airscout," 34. See also "New Instrument," 267.

69. Woodhouse, "Progress of Aviation," 1264.

70. Charles D. Rhodes, "The Cavalry of Today," *JUSCA* 24 (November 1913): 370.

71. Bell, "First Aeroplane," 1128–30.

72. Niemann, "Airships and Cavalry," 874–75. See also V. Sanden, "Cavalry and Aircraft in the Service of Reconnaissance," *JUSCA* 23 (March 1913): 829–31.

73. Riedl, "Cavalry in War," 290–91.

74. "Aeroplanes and the Cavalry," 124; Niemann, "Airships and Cavalry," 875.

75. Ezra B. Fuller Jr., "More Cavalry Instead of Less," *JUSCA* 22 (March 1912): 973–83. See also Ezra B. Fuller Jr., "Why We Need Cavalry," *JUSCA* 22 (March 1912): 966–69; and George B. Davis, "The Reorganization of the Cavalry," *JUSCA* 22 (March 1912): 797–805.

76. *Senate Committee on Military Affairs, Cavalry Regiments, 62nd Congress, 3rd Session 1912*, 1–2.

77. Hamilton S. Hawkins, "Cavalry," *JUSCA* 26 (April 1916): 591.

78. Hayne, "Organization and Employment," 202.

79. McClure, "Use of Cavalry," 965.

80. G. W. Moses, "Bulletin No. 18," *JUSCA* 24 (May 1914): 912.

81. A. L. Dade, "Reducing the Cavalry of the Regular Army," *JUSCA* 22 (May 1912): 1014–15. See also Lascelles, "Airship and Flying Machine," 212.

82. Rhodes, "Cavalry of Today," 369. See also Roscoe, "Effect of Aeroplanes," 857.

83. Rhodes, "Cavalry of Today," 369.

84. Rubottom, "Cavalry Reconnaissance," 25–26. See also Paul W. Beck, "Lecture on Aeroplanes" given before War College, Washington Barracks, D.C., November 17, 1911, Box 15, Folder 247, MHI.

85. G. W. Moses, "Communications and Reconnaissance on the Battlefield," *JUSCA* 23 (May 1913): 993–94; Moses, "Bulletin," 911. See also Niemann, "Airships and Cavalry," 874–5; and Hayne, "Organization and Employment," 202.

86. Olaf Schwarzkoff, "The Changed Status of the Horse in War," *JUSCA* 26 (January 1916): 347.

87. Roscoe, "Effect of Aeroplanes," 856.

88. Moses, "Communications and Reconnaissance," 996. See also Rhodes, "Cavalry of Today," 369.

89. F. B. Hennessy, "The Aviation Squadron in the Connecticut Maneuver Campaign," *JUSCA* 23 (November 1912): 458–76.

90. Ibid., 458.

91. "Editorial and Professional Notes: Military Aeroplanes," *Journal of the United States Artillery* 37 (January–February 1912): 102. See also Rhodes, "Cavalry of Today," 370.

92. Palmer, "Military Lessons," 2. See also "Bernhardi and French on Modern Cavalry," *Chicago Daily Tribune,* November 15, 1914, A4; "Squadron Fails in Mexico," *New York Tribune,* March 28, 1916, 3; and Reilly, "Aerial Service," 6.

93. Benjamin D. Foulois, "Military Aviation and Aeronautics," *Journal of the Military Service Institution of the United States* 52 (1913): 98–99,103–4.

94. "Army Air Scouts," 8.

95. "What Could Uncle Sam Do," B1.

96. Wisser, "Tactical and Strategical Use," 421.

97. Niemann, "Airships and Cavalry," 876–77. See also "Modern War," 965; Campbell, "Aeroplanes with Cavalry," 313; Rhodes, "Cavalry of Today," 370; James Parker, "The Value

of Cavalry as Part of Our Army," *JUSCA* 25 (July 1914): 8–9; and Hickok, "Role and Organization," 57.

98. Moses, "Communications and Reconnaissance," 994. See also Roscoe, "Effect of Aeroplanes," 857–58; Moses, "Bulletin," 912; Hawkins, "Cavalry," 601; Rhodes, "Cavalry of Today," 370.

99. J. A. Gaston, "Divisional Cavalry on the March and in Action," *JUSCA* 25 (April 1915): 601.

100. Moses, "Bulletin," 912.

101. Rhodes, "Cavalry of Today," 370.

102. Parker, "Value of Cavalry," 9.

103. "Aero Aids Maneuver," 11. See also "One Aeroplane," 26.

104. "Aero Aids Maneuver," 11.

105. "Making Aeroplanes Fighting Machines," *St. Louis Post-Dispatch*, October 11, 1914, B7.

106. "Aeroplanes Built," 8. See also "What Do You Know About the United States Army?" *Washington Post*, April 8, 1917, SM4.

107. W. H. Fauber, "U.S. Must Have Powerful Air Fleet," *New York Times*, February 13, 1916, X1. See also Henry J. Reilly, "Control of the Air Decides Battles," *Boston Daily Globe*, April 15, 1917, 16.

108. Reilly, "Control of the Air Decides Battles," 16.

109. "Air Attack on U.S. Peary's Prediction," *Boston Daily Globe*, November 25, 1917, 13.

110. "Cavalry's Last Days," *Kansas City Star*, June 18, 1917, 6.

111. Elbert E. Farman, "The Cavalry in the Present War," *JUSCA* 26 (April 1916): 625, 627–28.

112. Barrows, "Uhlans and Other Cavalry," 391, 398.

113. Henry J. Reilly, "Cavalry Branch Is a Big Factor in All War Moves: Army Expert Describes His Activities in Great European Conflict," *Chicago Daily Tribune*, October 20, 1915, 9.

114. Conventions vary, but in this work the official designation for units will be utilized when numbered abbreviations are used. For example, 2nd for Second Cavalry, 3rd for Third, etc.

115. Schwarzkoff, "Changed Status," 335–40.

116. Hawkins, "Cavalry," 581–89. See also Hickok, "Role and Organization," 73; A Cavalry Officer Abroad, "The Question of Organization," *JUSCA* 25 (October 1914): 205–7; One of Our Cavalry Officers, "Service with a French Cavalry Regiment," *JUSCA* 25 (October 1914): 232–33; A Cavalry Officer Abroad, "Report Upon Year Spent with French Cavalry," *JUSCA* 25 (January 1915): 447; and Henry J. Reilly, "Cavalry in the Great War," *JUSCA* 27 (April 1917): 478–79.

117. Hawkins, "Cavalry," 582.

118. Ibid., 583. See also Farman, "Cavalry in the Present War," 625.

119. Hawkins, "Cavalry," 584–85, 589.

120. Ibid., 601.

121. Henry J. Reilly, "Cavalry in Modern War," *JUSCA* 27 (November 1916): 294.

122. Ibid., 295.

123. Reilly, "Cavalry in the Great War," 482.

124. An Officer Abroad, "Notes on the European War," *JUSCA* 26 (July 1915): 54.

125. An Officer of High Rank, "What Has the World's War Taught Us Up to the Present Time That Is New in a Military Way," *JUSCA* 26 (July 1915): 97–98.

126. Ibid. See also "Dragoons Wrecked German Air Fleet: French War Office Describes Night Attacks of Cavalry on an Aeroplane Camp," *New York Times,* November 29, 1914, 2; Reilly, "Cavalry Branch Is a Big Factor," 9.

127. Allan R. Millett and Peter Maslowski, *For the Common Defense: A Military History of the United States of America,* rev. ed. (New York: Free Press, 1994), 337.

128. Ibid., 331.

129. Hofmann, *Through Mobility We Conquer,* 45; Millett and Maslowski, *For the Common Defense,* 333.

130. Johnson, *Wingless Eagle,* 139. For a more detailed treatment of airplane and cavalry use in the Punitive Expedition to Mexico, see chapter 4.

131. Morrow, *Great War in the Air,* 45.

132. Morrow's *Great War in the Air* provides a good treatment of European military aviation development during the war.

133. Timothy K. Nenninger, "American Military Effectiveness in the First World War," in *Military Effectiveness,* vol. 1., *The First World War,* ed. Allan R. Millett and Williamson Murray (Boston: Unwin Hyman 1988), 121.

134. Ibid., 139.

135. Stubbs and Connor, *Armor-Cavalry Part I,* 38; Hofmann, *Through Mobility We Conquer,* 2.

CHAPTER 3

1. *CJ* (UK) and *CJ* (US) (previously titled the *JUSCA*) resumed publication in April 1920.

2. Odom, *After the Trenches,* 64.

3. The following are some of the volumes discussing the cavalry in the war: Anglesey, *History of British Cavalry*; Cavalry School (U.S.) Academic Division, *Cavalry Operations during the World War* (Fort Riley, Kans.: Academic Division, The Cavalry School, 1929); David Kenyon, *Horsemen in No Man's Land: British Cavalry and Trench Warfare, 1914–1918* (Barnsley, South Yorkshire, UK: Pen and Sword, 2011); R. M. Preston, *The Desert Mounted Corps: An Account of the Cavalry Operations in Palestine and Syria, 1917–1918* (London: Constable, 1921); David R. Woodward, *Hell in the Holy Land: World War I in the Middle East* (Lexington: University Press of Kentucky, 2013); and Alexis Wrangel, *The End of Chivalry: The Last Great Cavalry Battles, 1914–1918* (New York: Hippocrene Books, 1982).

4. Rittgers, "Galloping Hooves to Rumbling Engines," 171–75.

5. United States Army American Expeditionary Forces, *Report of Superior Board on Organization and Tactics* (United States: s.n., 1919), 65, OCLC 18673488, MHI.

6. Odom, *After the Trenches,* 26–27.

7. George M. Russell, "Intelligence for Cavalry," *CJ* (US) 29 (October 1920): 254–59.

8. United Press, "Aircraft and Cavalry in Contest for Preference in Defense Planes," *Washington Post,* February 23, 1916, 2.

9. "The Race for Aeroplane Supremacy," *St. Louis Post-Dispatch,* February 27, 1916, B5; "What Do You Know About the United States Army?" *Washington Post,* April 8, 1917, SM4.

10. G. A. Weir, "Some Reflections on the Cavalry Campaign in Palestine," *Journal of the Royal United Service Institution* 67 (1922): 231.

11. Theo H. Thorne, "The Future of Cavalry: To the Editor of the Times," *Times* (London), April 22, 1919, 6.

12. Captain de la Vergne, "A Few Words about European Cavalry," *CJ* (US) 29 (July 1920): 155.

13. A. G. G., "Life and Politics," *Nation and the Athenaeum* 37 (April 1925): 41.

14. Elbridge Colby, "The Horse in War Today and Yesterday: I—Cavalry Still a Factor in Military Warfare," *Current History* 28 (1928): 446.

15. Joseph Cheries, "Army Service Experiences Questionnaire (1914–1921)" [hereafter "Army Questionnaire"], World War I Research Project, WWI 763, OCLC 457255624, MHI; Nelson Greer, "Army Questionnaire," WWI 3581, OCLC 457255624, MHI; Paul W. Krug, "Army Questionnaire," WWI 4289, OCLC 457255624, MHI.

16. Paul Weil, "Army Questionnaire," WWI 6063, OCLC 457255624, MHI.

17. Robert W. Jack, "Army Questionnaire," WWI 440, OCLC 457255624, MHI.

18. Ray A. Ingle, "Army Questionnaire," WWI 6149, OCLC 457255624, MHI; Ethan A. Nelson, "Army Questionnaire," WWI 1354, OCLC 457255624, MHI; Rea A. Nunnallee, "Army Questionnaire," WWI 5633, OCLC 457255624, MHI.

19. Mack Sheldon, "Army Questionnaire," WWI 4699, OCLC 457255624, MHI.

20. Francis Parsons, "Regular Cavalry," *Life,* January 1, 1920, 42.

21. "Editorial Comment: Cavalry Journal Reappears," *CJ* (US) 29 (April 1920): 81.

22. Ibid., 82.

23. "Editorial Comment: A Cavalry Army," *CJ* (US) 30 (October 1921): 420.

24. George S. Patton Jr., "What the World War Did for Cavalry," *CJ* (US) 31 (April 1922): 168.

25. "By other editors: Cavalry Still Useful," *Christian Science Monitor,* September 19, 1918, 6.

26. John J. Pershing, "A Message to the Cavalry," *CJ* (US) 29 (April 1920): 5–6.

27. William C. Sherman, "Cavalry and Aircraft," *CJ* (US) 30 (January 1921): 26.

28. Grover O'Neill, Diary Entry June 29, 1918, June 24–July 18, 1918, IRIS 00121421, Box 167.606-1-167.607-5 1917–1918, AFHRA.

29. F. F. Hunter, "Report of Air Service, 1st Army A.E.F., France, November 16, 1918," Call No. 167.401-21, IRIS 00120736 in vol. 9, 1909–1939, AFHRA.

30. Earl Haig, "Introductory Remarks," *CJ* (UK) 10 (April 1920): 5–6.

31. *Memorandum on Army Training Collective Training Period, 1927* (London: HMSO, 1927), December 15, 1927, 8, 23, WO 231/210, TNA.

32. See Appendix A for the changing capabilities of airplanes from 1908 to 1929.

33. Staff College Committee No. 7, Monograph on Air Service, 1919–1920, Call No. 167.601-5, 1918–1920, IRIS 00121313, Box 167.601-2-167.601-13 1917–1918, AFHRA.

34. For additional discussions on aerial radio and photography, see Paul W. Beck, "Lecture on Aeroplanes," November 17, 1911, OCLC 63674200, MHI; "American-Developed Radio Telephone Success in Airplanes," *Telephony,* November 23, 1918, 17, http://

www.earlyradiohistory.us/1918air.htm; "Wilson Directs Air Maneuvers by Radiophone," (Boise) *Idaho Statesman,* November 23, 1919, 2; A. J. Tittinger, "The Future of Cavalry," *CJ* (US) 29 (April 1920): 68; Russell, "Intelligence for Cavalry," 258; N. N. Golovine, "Cavalry Reconnaissance: The Modern Service of Information and Cavalry's Role in It," *CJ* (US) 31 (April 1922): 193; Edward Davis, "The British Cavalry in Palestine and Syria," pt. 4, *CJ* (US) 31 (January 1923): 64; Edward M. Fickett, "A Study of the Relationship between the Cavalry and the Air Service in Reconnaissance," *CJ* (US) 32 (October 1923): 419; T. L. Leigh-Mallory, "Co-operation of Aircraft with Cavalry," *CJ* (UK) 17 (January–October 1927): 280; Edward J. Dwan, "The Reorganization of the United States Cavalry," *CJ* (UK) 19 (January–October 1929): 614; Samuel B. Fishbein, *Edward Steichen: A Brief Biographical Sketch of This Pioneer and the History of Photographic Technology Prior to and during World War I*, National Air and Space Museum Smithsonian Institution, 1994; George C. Larson, "Moments and Milestones: Can You Hear Me Now? When Radio Communication Took to the Air," *Air & Space Magazine* (March 2011), 80, http://www.airspacemag.com/history-of-flight/moments-and-milestones-can-you-hear-me-now-10947/?no-ist.

35. Millett and Maslowski, *For the Common Defense,* 384–86; "New Bill for Army of Three Branches," *St. Louis Post-Dispatch,* January 4, 1920, A1; "Divides New Army into 3 Branches," *Washington Post,* January 4, 1920, 10; and "For Three-Branch Army," *Baltimore Sun,* January 4, 1920, 2.

36. "Editorial Comment: Cavalry Journal Reappears," 81.

37. "Willard Ames Holbrook," Arlington National Cemetery, last modified May 20, 2006, http://www.arlingtoncemetery.net/waholbrook.htm.

38. Willard A. Holbrook, "A Few Words to the Cavalry," *CJ* (US) 29 (October 1920): 249–50.

39. Cavalry School Department of Tactics, *Employment of Cavalry, 1924–1925* (Fort Riley, Kans., 1925), OCLC 13168234, MHI.

40. Ibid., 7.

41. Weigley, *History of the United States Army,* 328; Leonard Wood, "Cavalry's Role in the Reorganization," *CJ* (US) 29 (July 1920): 113.

42. M. N. MacLeod, "Survey Work in Modern War," *CJ* (UK) 11 (January–October 1921): 169.

43. Paul Davison, "The Cavalry's Cavalry: A Study of Armored Cars, Their Powers and Limitations, and Some Notes as to Their Tactical Employment," *CJ* (US) 36 (January 1927): 92.

44. Leonard Nason, "Horse and Machine," *CJ* (US) 38 (April 1929): 195. See also A. W. H. James, "Co-operation of Aircraft with Cavalry: General Principles," *CJ* (UK) 10 (January–October 1920): 135; Hamilton S. Hawkins, "The Role of Cavalry," *CJ* (US) 29 (October 1920): 262; "Fundamentals of Cavalry Training Policy: Cavalry Memorandum No. I War Department, December 10, 1920," *CJ* (US) 30 (April 1921): 183; MacLeod, "Survey Work," 169; A. R. Mulliner, "Cavalry Still an Essential Arm," *CJ* (UK) 17 (January–October 1927): 641; H. V. S. Charrington, "Where Cavalry Stands To-day," *CJ* (UK) 17 (January–October 1927): 430; George Barrow, "The Future of Cavalry," *CJ* (UK) 19 (January–October 1929): 179–80; James Parker, "The Cavalryman and the Rifle," *CJ* (US) 37 (July 1928): 367.

45. Cavalry School, *Employment of Cavalry,* 7.

46. Stacy Hinkle, "Wings and Saddles: The Air and Cavalry Punitive Expedition of 1919," *Southwestern Studies* 5 (1967): 5.

47. MacLeod, "Survey Work," 169.

48. "Reconnaissance," *CJ* (US) 30 (October 1921): 353.

49. Jack Wade, "Lest Ye Forget," *CJ* (US) 31 (July 1922): 312.

50. W. D. Croft, "Notes on Armoured Cars," *CJ* (UK) 15 (January–October 1925): 162. See also Hawkins, "Role of Cavalry," 262; "Fundamentals of Cavalry," 183; von Ammon, "Cavalry Lessons of the Great War from German Sources," *CJ* (US) 30 (October 1921): 358; Ernest N. Harmon, "The Second Cavalry in the Meuse-Argonne," *CJ* (US) 31 (January 1922): 17; Patton, "What the World War," 168–69; Hamilton S. Hawkins, "The Importance of Modern Cavalry and its Role as Affected by Developments in Airplane and Tank Warfare," *CJ* (US) 35 (October 1926): 489; Charrington, "Where Cavalry Stands," 430; Parker, "Cavalryman and the Rifle," 367.

51. Cavalry School, *Employment of Cavalry,* 7.

52. Wood, "Cavalry's Role," 113. See also "Fundamentals of Cavalry," 180–85; "Topics of the Day: The Expedition into Mexico to Recover the Pierson Airplane," *CJ* (US) 30 (July 1921): 305–6; Harmon, "Second Cavalry in the Meuse-Argonne," 15–16; Hawkins, "Importance of Modern Cavalry," 489; Davison, "Cavalry's Cavalry," 90–95; Charrington, "Where Cavalry Stands," 430; Mulliner, "Cavalry Still Essential," 641–42; E. G. Hume, "Some Thoughts on Modern Reconnaissance," *CJ* (UK) 18 (January–October 1928): 230; "The Future of Cavalry: A Lecture at the University of Bristol on 7th February, 1929," *CJ* (UK) 19 (January–October 1929): 378–79; C. C. Benson, "Mechanization—Aloft and Alow," *CJ* (US) 38 (January 1929): 60.

53. Hinkle, "Wings and Saddles," 6, 23.

54. Hawkins, "Role of Cavalry," 262.

55. Wade, "Lest Ye Forget," 312.

56. Leo G. Hefferman, "Co-operation between Cavalry and Air Service." *CJ* (US) 34 (April 1925): 151; Sherman, "Cavalry and Aircraft," 28. See also Leigh-Mallory, "Cooperation of Aircraft," 279–80.

57. Cavalry School, *Employment of Cavalry,* 8, 15.

58. "Cavalry Staff Exercise Oxford April 19–22, 1927," 16, WO 279/58, TNA; Hawkins, "Importance of Modern Cavalry," 489. See also Charrington, "Where Cavalry Stands," 419–30 for more on the cavalry's ability to question inhabitants, take prisoners, and examine documents. American, British, and French writers stated the weaknesses of gathering information exclusively from the air. See James, "Co-operation of Aircraft," 137; "Foreign Magazines," *CJ* (UK) 18 (January–October 1928): 664–68.

59. Hinkle, "Wings and Saddles," 32.

60. Croft, "Notes on Armoured Cars," 163.

61. Mulliner, "Cavalry Still Essential," 641; "Cavalry Staff Exercise," 16.

62. Fickett, "Study of the Relationship," 414; Cavalry School, *Employment of Cavalry,* 7–8.

63. U.S. War Department, *FSR, United States Army, 1923* (Washington, D.C.: GPO, 1924), 32. (Hereafter *FSR 1923.*)

64. "Future of Cavalry: A Lecture," 378.

65. Leigh-Mallory, "Co-operation of Aircraft," 280.

66. U.S. Army Command and General Staff School, *Tactical Principles and Decisions: Volume 1 Marches, Halts, and Reconnaissance and Security,* 1920, rev. ed. (Fort Leavenworth, Kans.: General Service Schools Press, 1922), 101. See also Fickett, "Study of the Relationship," 420; C. B. Dashwood Strettell, "Cavalry in Open Warfare, Illustrated by the Operations Leading Up to the Occupation of Mosul in November, 1918," *Journal of the Royal United Service Institution* 66 (1921): 617; Colby, "Horse in War," 446–47.

67. Cavalry School, *Employment of Cavalry,* 8.

68. James, "Co-operation of Aircraft," 135. See also Hamilton S. Hawkins, "German Ideas on Modern Cavalry," *CJ (US)* 31 (April 1925): 155–68.

69. Mulliner, "Cavalry Still an Essential Arm," 641.

70. Stacy C. Hinkle, *Wings over the Border: The Army Air Service Armed Patrol of the United States-Mexico Border, 1919–1921* (El Paso: Western Press, University of Texas, 1970), 12.

71. "Fundamentals of Cavalry," 183.

72. U.S. Army Command and General Staff College, *Tactical Principles,* 101.

73. Barrows, "Future of Cavalry," 179.

74. *FSR 1923,* 21.

75. Sherman, "Cavalry and Aircraft," 29–30.

76. Hinkle, *Wings over the Border,* 2, 7, 9.

77. "Editorial Comment: Cavalry Journal Reappears," 81; Weir, "Some Reflections," 231–32.

78. Weir, "Some Reflections," 231–32.

79. Colby, "Horse in War," 446.

80. "Dominion and Foreign Magazines," *CJ* (UK) 15 (January–October 1925): 96.

81. A Dweller on Earth, "A Little Bird Told Me," *CJ* (UK) 18 (January–October 1928): 603.

82. For an example of the rare exception, see "The Action of the Future," *JUSCA* 19 (January 1909): 635.

83. Fenton Jacobs, "The Cavalry," *CJ* (US) 30 (April 1921): 124–29.

84. Weir, "Some Reflections," 233.

85. Ibid., 232.

86. "Cavalry, an Essential Arm," *CJ* (UK) 12 (January 1922): 6.

87. "The Horse in War," *CJ* (US) 34 (October 1925): 412.

88. Hawkins, "Importance of Modern Cavalry," 487.

89. *Memorandum on Army Training,* 8., *1927* .

90. Parker, "Cavalryman and the Rifle," 367.

91. K. S. Bradford, "Modern Cavalry," *Infantry Journal* 35 (December 1929): 564, 566.

92. "The Cavalry Spirit," *CJ* (UK) 12 (January–October 1922): 225.

93. Basil Henry Liddell Hart, "After Cavalry—What?" *The Atlantic Monthly* 136 (September 1925): 418.

94. Minutes of the Second Meeting of the Sub-committee on the Strength and Organisation of the Cavalry, December 19, 1927, 21, CAB 16/77, TNA.

95. Minutes of the Third Meeting of the Sub-committee on the Strength and Organisation of the Cavalry, February 9, 1928, 8, CAB 16/77, TNA.

96. "Editorial Comment: Modern Defense Policies Increase Importance of Cavalry," *CJ* (US) 32 (October 1923): 450.

97. James, "Co-operation of Aircraft," 134. See also Tittinger, "Future of Cavalry," 68.

98. Fickett, "Study of the Relationship," 414.

99. Dweller on Earth, "Little Bird," 601–4.

100. Ibid., 603.

101. Ibid., 604.

102. H. D. Fanshane, "Cavalry in Mesopotamia in 1918," *CJ* (UK) 10 (January–October 1920): 429; "Mechanical Warfare," *CJ* (UK) 12 (January–October 1922): 156; Rex Osborne, "Operations of the Mounted Troops of the Egyptian Expeditionary Force," *CJ* (UK) 13 (January–October 1923): 24; de la Vergne, "Few Words," 157; Edward Davis, "The British Cavalry in Palestine and Syria," pt. 3, *CJ* (US) 31 (October 1922): 398; Golovine, "Cavalry Reconnaissance," 186; E. G. Hume, "Notes on Modern French Cavalry," *CJ* (UK) 15 (January–October 1925): 213; "Topics of the Day: Air Service and Other Auxiliaries for Cavalry," *CJ* (US) 31 (October 1922): 412; Hawkins, "Importance of Modern Cavalry," 489; "Topics of the Day: Reorganizing of the British Cavalry," *CJ* (US) 36 (July 1927): 480; Leigh-Mallory, "Cooperation of Aircraft," 279; Charrington, "Where Cavalry Stands," 422; Davison, "Cavalry's Cavalry," 90; H. H. Arnold, "The Cavalry-Air Corps Team," *CJ* (US) 37 (January 1928): 76; Dweller on Earth, "A Little Bird," 604.

103. Tittinger, "Future of Cavalry," 68. See also "One of the Faithful" [pseud.], "Faith in and a Doctrine for the Cavalry Service," *CJ* (US) 36 (April 1927): 227–31 for the argument that the cavalry benefited most from aviation.

104. E. H. H. Allenby, "Cavalry in Future Wars," *CJ* (US) 30 (January 1921): 2. See also Allenby's comments as reported in "'Cavalry Journal' Committee," *CJ* (UK) 16 (1926): 3.

105. Weir, "Some Reflections," 234; and "Topics of the Day: Cavalry Discussed by Royal United Service Institution," *CJ* (US) 31 (April 1922): 202. For additional arguments discussing the added efficiency provided to the cavalry by mechanical means, see Kirby Walker, "Cavalry in the World War," *CJ* (US) 33 (January 1924): 11–22; Kenyon A. Joyce, "The British Army Maneuvers, 1925," *CJ* (US) 35 (January 1926): 17–20; For the strength and importance of cooperation, see Davison, "Cavalry's Cavalry," 90–95.

106. "One of the Faithful" [pseud.], "Faith in and a Doctrine," 227. See also Herbert B. Crosby, "Cavalry of Today," *CJ* (US) 36 (October 1927): Frontispiece; N. N. Golovine, "Cavalry on the Flank and in the Rear," *CJ* (US) 31 (January 1922): 40–59; N. N. Golovine, "Modern Cavalry and Its Organization," *CJ* (US) 35 (January 1926): 30; "Topics of the Day: Two More Types of Guns for Cavalry," *CJ* (US) 37 (July 1928): 437; Allenby, "Cavalry in Future Wars," 2; General Brechard, "The French Cavalry," *CJ* (US) 38 (October 1929): 486.

107. Fanshane, "Cavalry in Mesopotamia," 429; Osborne, "Operations of the Mounted Troops," 24.

108. Tittinger, "Future of Cavalry," 68. See also Mulliner, "Cavalry Still an Essential Arm," 641; and Hume, "Some Thoughts," 211–32 for more information about increasing the radius of cavalry operations by using technology.

109. George E. Mitchell, "The Rout of the Turks by Allenby's Cavalry (Continued)," *CJ* (US) 29 (July 1920): 195. See also Davis, "British Cavalry in Palestine and Syria," pt. 3,

395–402. There were a few cases where the arguments mirrored those prior to the war, such as Hawkins, "Importance of Modern Cavalry," 487–99. He argued vaguely that reconnaissance by aviation could save the cavalry from exhausting work. He argued after the war that aviation could support the cavalry.

110. J. R. H. Cruikshank, "How the Cavalry Exploits a Victory: Being Extracts from the Diary of a Subaltern under Allenby in Palestine," *CJ* (US) 32 (April 1923): 164.

111. Mitchell, "Rout of the Turks," 203. See also Edward Davis, "The British Cavalry in Palestine and Syria," pt. 1, *CJ* (US) 31 (April 1922): 123–29.

112. Ernest N. Harmon, "The Second Cavalry in the St. Mihiel Offensive," *CJ* (US) 30 (July 1921): 282–89. See also Harmon, "Second Cavalry in the Meuse-Argonne," 10–18.

113. Redding F. Perry, "The 2d Cavalry in France," *CJ* (US) 37 (January 1928): 29.

114. Pershing, "Message to the Cavalry," 5–6. See also Harmon, "Second Cavalry in the St. Mihiel," 282–89.

115. James, "Co-operation of Aircraft," 133.

116. Tittinger, "Future of Cavalry," 68; Hume, "Notes on Modern French Cavalry," 33. See also Hawkins, "German Ideas," 155–68.

117. "Topics of the Day: Air Service and Other Auxiliaries," 412. See also Hume, "Notes on Modern French Cavalry," 27–38 for references to the importance of combined training.

118. James, "Co-operation of Aircraft," 133.

119. LeRoy Eltinge, "Review of Our Cavalry Situation," *CJ* (US) 29 (April 1920): 14–22.

120. Fickett, "Study of the Relationship," 420.

121. Ibid., 412–22. See also James, "Co-operation of Aircraft," 133–41; "Mechanical Warfare," 156.

122. Hinkle, "Wings and Saddles," 3–4.

123. Kenneth Baxter Ragsdale, *Wings over the Mexican Border: Pioneer Aviation in the Big Bend* (Austin: University of Texas Press, 1984), 74–75; Morrow, *Great War in the Air*, 339; Lucien H. Thayer, *America's First Eagles: The Official History of the U.S. Air Service, A.E.F. (1917–1918)* (San Jose, Calif. and Mesa, Ariz.: Bender and Champlin Fighter Aces Museum Press, 1983), 243–44.

124. "Ready for War in the Big Bend of Texas: Airmen and Cavalrymen Maintain Constant Vigilant Patrol along Section of Border Often Overrun by Mexicans," *New York Times*, January 25, 1920, 42; Ragsdale, *Wings over the Mexican Border*, 74.

125. Hinkle, *Wings over the Border*, 11.

126. Hinkle, "Wings and Saddles," 4–5.

127. Ibid., 39–40.

128. "Topics of the Day: Expedition into Mexico," 305.

129. Ibid., 305–6.

130. Sherman, "Cavalry and Aircraft," 29.

131. Ibid., 63–64.

132. Hinkle, "Wings and Saddles," 33, 35.

133. W. P. King, "The Fifteen Days' Training Period of the 62d Cavalry Division, Camp Meade, 1922," *CJ* (US) 31 (January 1923): 72.

134. Adna R. Chaffee, "The Maneuvers of the First Cavalry Division, September–October 1923," *CJ* (US) 33 (April 1924): 135, 140, 152–53, 156. See also Hefferman, "Cooperation," 151.

135. Ernest Hinds, *Report Air-Ground Maneuvers, San Antonio, Texas, May 15–21, 1927* (Fort Sam Houston, Tex.: Headquarters, Eighth Corps Area, 1927), OCLC 54074817, MHI.

136. R. E. Cummins to Chief Umpire, "Umpire's Report, Army Maneuvers," May 19, 1927, in Hinds, *Report, Air-Ground Maneuvers.*

137. B. G. Chynoweth to Adjutant General, "Report of Press Relations in the Maneuvers," May 21, 1927, in Hinds, *Report, Air-Ground Maneuvers,* 4.

138. George Dillman, "Joint Cavalry and Air Corps Training," *CJ* (US) 36 (October 1927): 560.

139. Gordon Swanborough and Peter M. Bowers, *United States Military Aircraft since 1909* (Washington, D.C.: Smithsonian Institution Press, 1989), 212.

140. Dillman, "Joint Cavalry," 560–67.

141. George Dillman, "1st Cavalry Division Maneuvers," *CJ* (US) 37 (January 1928): 48, 65.

142. Ibid., 54.

143. Benson, "Mechanization," 58.

144. United States Army Air Service, *Annual Report of the Chief of Air Service for the Fiscal Year Ending June 30, 1922,* (Washington, D.C.: GPO 1922), OCLC 23822134, MHI.

145. Hinkle, *Wings over the Border,* 40.

146. U.S. Army American Expeditionary Forces, *Report of Superior Board,* 65, 72.

147. Eltinge, "Review of Our Cavalry Situation," 14–22; Contributed, "Cavalry Organization," *CJ* (US) 29 (July 1920): 121; "Topics of the Day: Air Service and Other Auxiliaries for Cavalry," 412; Dillman, "Joint Cavalry," 560.

148. Mitchell, "Rout of the Turks," 195.

149. Davis, "British Cavalry in Palestine and Syria," pt. 3, 395–402.

150. "Topics of the Day: Air Service and Other Auxiliaries for Cavalry," 412.

151. Crosby, "Cavalry of Today," Frontispiece.

152. "Topics of the Day: Two More Types of Guns for Cavalry," 437.

153. Ibid. See also Benson, "Mechanization," 58.

154. "Topics of the Day: Extracts from the Annual Report of the Chief of Cavalry," *CJ* (US) 37 (January 1928): 108.

155. See William Mitchell, *Winged Defense: The Development and Possibilities of Modern Air Power—Economic and Military* (1925; repr., New York: Dover, 1988), 20–22, 225.

156. Smith, *British Air Strategy,* 15–19.

157. *Memorandum on Training Carried Out during the Collective Training Period, 1925,* (London: HMSO, 1925), 5, WO 231/207, TNA.

158. *Memorandum on Army Training Collective Training Period, 1926* (London: HMSO, 1926), 12, WO 231/208, TNA.

159. *Memorandum on Training, 1927.*

160. Ibid.

161. "Cavalry Staff Exercise."

162. From H. J. Creedy at the WO for the information of Lord Colwyn's Committee, "The Army and Royal Air Force," November 27, 1925, WO 32/2782, TNA.

163. Minutes of the Fourth Meeting of the Subcommittee on the Strength and Organisation of the Cavalry, February 21, 1928, Evidence of Sir Hugh Trenchard, 9, CAB 16/77, TNA.

164. Ibid.

165. Mitchell, "Rout of the Turks," 183. See also T. Miller Maguire, "The Cavalry Career of Field Marshall Viscount Allenby," *CJ* (UK) 10 (January–October 1920): 379–84; Patton, "What the World War," 165–72; "Topics of the Day: Cavalry Discussed," 201–2; Cruikshank, "How the Cavalry Exploits a Victory," 163–70; Stephen H. Sherrill, "The Experiences of the First American Troop of Cavalry to Get into Action in the World War," *CJ* (US) 32 (April 1923): 153–59; Edward Davis, "The British Cavalry in Palestine and Syria: Conquest of Syria (continued)," pt. 5, *CJ* (US) 32 (October 1923): 435–44.

166. James, "Co-operation of Aircraft," 243.

167. Harmon, "Second Cavalry in the St. Mihiel Offensive," 283. See also Harmon, "Second Cavalry in the Meuse-Argonne," 10–18.

168. Weir, "Some Reflections," 230.

169. Patton, "What the World War," 170. For additional discussion of protection from aerial attack, see E.M. Whiting, "Protection from Enemy Aircraft," *CJ* (US) 37 (January 1928): 103–6.

170. Davis, "British Cavalry in Palestine and Syria," pt. 4, 62.

171. *Memorandum on Training, 1925*, 10.

172. C. F. Houghton, "Attack Aviation vs. Cavalry," *CJ* (US) 38 (April 1929): 224.

173. WO, *FSR, Volume II: Operations, 1929* (London: HMSO, 1929), 14.

174. Houghton, "Attack Aviation vs. Cavalry," 222.

175. *FSR 1923*, 3.

176. Ibid., 11.

177. WO, *FSR, Volume II: Operations, 1924* (London: HMSO, 1924), 12.

178. Ibid.

179. WO, *Cavalry Training, Volume II: War, 1929* (London: HMSO, 1929), 21; *FSR 1923*, 32.

180. WO, *Cavalry Training Volume II: War 1929*, 21–22.

181. Hart, "After Cavalry," 414.

182. Fickett, "Study of the Relationship," 415–18. See Appendix B for reproduced tables.

183. *FSR 1923*, 33.

184. WO, *Cavalry Training, Volume II: War, 1929*, 22.

185. Arnold, "Cavalry-Air Corps Team," 76.

186. U.S. Army Command and General Staff College, *Tactical Principles and Decisions*, 103.

187. *FSR 1923*, 104.

188. WO, *Cavalry Training, Volume II: War, 1929*, 12.

189. Ibid., 16.

190. WO, *FSR, Volume II: Operations, 1929*, 10.

CHAPTER 4

1. For detailed studies of the success and failure of air policing, see Corum and Johnson, *Airpower in Small Wars*; Omissi, *Air Power*; Brian Robson, *Crisis on the Frontier: The Third Afghan War and the Campaign in Waziristan, 1919–20* (Staplehurst: Spellmount, 2007); Andrew Roe, *Waging War in Waziristan: The British Struggle in the Land of Bin Laden, 1849–1947* (Lawrence: University Press of Kansas, 2010); Smith, *British Air Strategy*; Philip Anthony Towle, *Pilots and Rebels: The Use of Aircraft in Unconventional Warfare, 1918–1988* (London: Brassy's, 1989); Lawrence James, *Imperial Rearguard: Wars of Empire, 1919–1985* (London, Brassey's Defence, 1998); Keith Jeffery, *British Army and the Crisis of Empire, 1918–22* (Manchester, UK: Manchester University Press, 1984).

2. Edward J. Stackpole Jr., "The National Guard Cavalry," *CJ* (US) 47 (March–April 1938): 101.

3. Badsey, *Doctrine and Reform in the British Cavalry*, 303–4.

4. Ibid.

5. Strettell, "Cavalry in Open Warfare," 598.

6. Ibid., 600.

7. Mulliner, "Cavalry Still an Essential Arm," 640.

8. Ibid., 647.

9. Ibid.

10. Barrow, "The Future of Cavalry," 178.

11. Ibid., 184.

12. "The Future of Cavalry: A Lecture," 376.

13. Minutes of the Second Meeting of the Sub-committee on the Strength and Organisation of the Cavalry, December 19, 1927, 7, CAB 16/77, TNA.

14. Minutes of the Third Meeting of the Sub-committee on the Strength and Organisation of the Cavalry, 8.

15. Minutes of the Fourth Meeting of the Sub-committee on the Strength and Organisation of the Cavalry, Evidence of Colonel S. C. Peck," February 21, 1928, 2, CAB 16/77, TNA.

16. "Committee of Imperial Defence: Sub-committee on the Strength and Organisation of the Cavalry Report," May 3, 1928, 6, CAB 16/77, TNA.

17. "Appendix II: Memoranda by the War Office on the Report of the Geddes Committee, Paper A-General Staff Paper Circulated to the Cabinet by the Secretary of State for War," January 10, 1922, 2–5, CAB 27/164, TNA.

18. Henry Higgs, "The Geddes Report and the Budget," *Economic Journal* 32, no. 126 (June 1922): 251.

19. See table 3 for selected committees and members.

20. Smith, *British Air Strategy*, 22, 31, 121.

21. "Interim Report of the Committee on National Expenditure," December 14, 1921, 6–7, T 172/1228 Part 15, TNA. See also Smith, *British Air Strategy*, 15–19.

22. Smith, *British Air Strategy*, 15–16.

23. "Committee on National Expenditure Interim Report," December 14, 1921, 10, AIR 8/41, TNA; Omissi, *Air Power*, 38.

24. Jeffery, *British Army*, 67.

25. Memorandum by the Chief of the Air Staff on Air Power Requirements of the Empire, December 9, 1918, 4, CAB 24/71, TNA.

26. Ibid., 12.

27. Corum and Johnson, *Airpower in Small Wars,* 52.

28. For detailed studies of the success and failures of air policing, see Corum and Johnson, *Airpower in Small Wars;* Omissi, *Air Power;* Robson, *Crisis on the Frontier;* Roe, *Waging War in Waziristan;* Smith, *British Air Strategy;* and Towle, *Pilots and Rebels.*

29. Colwyn Committee on Navy, Army, and Air Force Expenditure, January 1, 1925–December 31, 1925, AIR 19/121, TNA.

30. A. E. Borton, "The Use of Aircraft in Small Wars," *Journal of the Royal United Service Institution* 65 (1921): 317–18. See also Minutes of the First Meeting of the Sub-committee on the Strength and Organisation of the Cavalry, December 8, 1927, 24, CAB 16/77, TNA; Minutes of the Third Meeting of the Sub-committee on the Strength and Organisation of the Cavalry, 8; Evidence of Sir Hugh Trenchard, 10–13.

31. Omissi, *Air Power,* ix.

32. Higgs, "Geddes Report," 254, 257.

33. "Interim Report of the Committee on National Expenditure," n.d., 8–9, AIR 8/41, TNA.

34. Minutes of the Fourth Meeting of the Committee Appointed to Examine Part I (Defence Departments) of the Report of the Geddes Committee on National Expenditure, January 23, 1922, 1–2, CAB 27/164, TNA.

35. Roe, *Waging War in Waziristan,* 130. See also Robson, *Crisis on the Frontier.*

36. "Interim Report of the Committee on National Expenditure," December 14, 1921, 7.

37. Ibid.

38. Ibid., 40, 42.

39. Appendix II: Memoranda by the War Office, 7, CAB 27/164, TNA.

40. Ibid.

41. Winston S. Churchill, "Report of the Committee Appointed to Examine Part I (Defence Departments) of the Report of the Geddes Committee on National Expenditure," February 4, 1922, 3, 10–11, CAB 27/164, TNA.

42. Ibid., 7–12.

43. Colwyn Committee: Note upon Alleged Duplication of Ground Services by the Army and Air Force, November 9, 1925, AIR 19/121, TNA.

44. "Committee on Navy, Army and Air Force Expenditure Report," December 23, 1925, 4, AIR 19/122, TNA.

45. "Standing Committee on Expenditure Report of the Colwyn Committee, Memorandum by the Secretary of State for Air," n.d. [1925?], AIR 19/120, TNA.

46. To Stanley Baldwin, General Remarks on Defence Expenditure, Report, Committee on Navy, Army and Air Force Expenditure, December 25, 1925, 4, AIR 19/122, TNA.

47. "The Future of Cavalry, Views of Field-Marshalls Viscount Allenby and Sir William Robertson, Enclosure No. 2," November 3, 1927, 15-17, CAB 16/77, TNA.

48. Further Memorandum by the Secretary of State for War, Army Estimates 1928, Committee of Imperial Defence: Sub-committee on the Strength and Organisation of

the Cavalry, The Cabinet Committee on Expenditure, November 11, 1927, 2–3, CAB 16/77, TNA.

49. Minutes of the First Meeting of the Sub-committee on the Strength and Organisation of the Cavalry, 2–5.

50. Ibid., 7.

51. Ibid., 9.

52. Ibid., 9–10.

53. Minutes of the First Meeting of the Sub-committee on the Strength and Organisation of the Cavalry, 24.

54. Minutes of the Third Meeting of the Sub-committee on the Strength and Organisation of the Cavalry, 16–17.

55. Minutes of the Fifth Meeting of the Sub-committee on the Strength and Organisation of the Cavalry, March 18, 1928, 2, CAB 16/77, TNA.

56. Omissi, *Air Power,* 13.

57. See Roe, *Waging War in Waziristan*; Omissi, *Air Power*; James, *Imperial Rearguard*; Corum and Johnson, *Airpower in Small Wars*; Jeffery, *British Army*; and Towle, *Pilots and Rebels.*

58. Corum and Johnson, *Airpower in Small Wars,* 5. See also James, *Imperial Rearguard,* 51, 71.

59. Corum and Johnson, *Airpower in Small Wars,* 5, 61.

60. Roe, *Waging War in Waziristan,* 131.

61. Ibid., 63.

62. Omissi, *Air Power,* 60.

63. Suggested Heads of Report, Committee of Imperial Defence: Sub-committee on the Strength and Organisation of the Cavalry, March 7, 1928, 7, CAB 16/77, TNA.

64. Towle, *Pilots and Rebels,* 18.

65. Omissi, *Air Power,* 9.

66. Ibid., 12.

67. Osborne, "Operations of the Mounted Troops," 24–31.

68. Rif refers to the frontier of Morocco.

69. Charrington, "Where Cavalry Stands," 420–21.

70. Roe, *Waging War in Waziristan,* 130, 134, 141.

71. Ibid.

72. Towle, *Pilots and Rebels,* 12.

73. Roe, *Waging War in Waziristan,* 23.

74. Corum and Johnson, *Airpower in Small Wars,* 56.

75. Ibid., 81; Roe, *Waging War in Waziristan,* 136.

76. Omissi, *Air Power,* 29.

77. Ibid., 70–71.

78. Richard R. Muller, "Close Air Support: The German, British and American Experiences, 1918–1941," in *Military Innovation in the Interwar Period,* ed. Williamson Murray and Allan R. Millett (Cambridge: Cambridge University Press, 1996), 172.

79. Omissi, *Air Power,* 16, 27.

80. Telegram from the Secretary of State for the Colonies to the Officer Administering the Government of Palestine, January 18, 1926, T 161/262, TNA.

81. Unsigned draft letter to Undersecretary of State, August 1925, T 161/262, TNA; Letter on economy in Palestine and Transjordania, February 7, 1926, T 161/262, TNA.

82. Weigley, *History of the United States Army*, 400–401.

83. "Airplanes Used by Army Officers for Huge Bonfire: Witnesses Tell Congress Probers How American Officers Ordered Burning of Planes Burned near Toul, France," *Atlanta Constitution*, July 31, 1919, 2; "Army Charged with Burning 150 Airplanes: Former Soldiers Tell House Investigators of 'Million-Dollar Bonfire' in June at French Flying Field," *New York Tribune*, July 31, 1919, 2; "Sickening Administrative Inefficiency," *Los Angeles Times*, August 2, 1919, II4."Topics of the Times: Destruction Really an Economy," *New York Times*, August 7, 1919, 6; "Seek Graft in Aircraft: Committee of Congress Starts Hearings Today in Matter of Spruce Production," *Los Angeles Times*, August 11, 1919, 13.

84. "Lodge Describes War Waste at Big Rally in Brooklyn: Administration Purchases Revealed as Riot of Extravagance without Result," *New York Tribune*, October 19, 1920, 1.

85. Brian McAllister Linn, *The Echo of Battle: The Army's Way of War* (Cambridge, Mass.: Harvard University Press, 2007), 118–19.

86. Alfred F. Hurley, *Billy Mitchell: Crusader for Air Power* (New York: Franklin Watts, 1964), 101.

87. William Mitchell, *Our Air Force: The Keystone of National Defense* (New York: E. P. Dutton, 1921), 2.

88. Hurley, *Billy Mitchell*, 57.

89. Charles M. Melhorn, *Two-Block Fox: The Rise of the Aircraft Carrier, 1911–1929*, (Annapolis, Md.: Naval Institute Press, 1974), 67–71; Hurley, *Billy Mitchell*, 59.

90. Melhorn, *Two-Block Fox*, 71; William A. Moffett, "Some Aviation Fundamentals," *United States Naval Institute Proceedings* 51 (October 1925): 1873.

91. Melhorn, *Two-Block Fox*, 70.

92. Moffett, "Some Aviation Fundamentals," 1873.

93. William Mitchell, *Memoirs of World War I: From Start to Finish of Our Greatest War* (New York: Random House, 1960), 4.

94. Ibid., 5.

95. Linn, *Echo of Battle*, 116–23.

96. *Army Air Service Hearing on H.R. 10827, Before the Senate Committee on Military Affairs*, Sixty-Ninth Cong. 42 (1926).

97. Corum and Johnson, *Airpower in Small Wars*, 11–16.

98. Ibid., 17–19.

99. Hinkle, *Wings over the Border*, 39; "Ready for War in the Big Bend of Texas," 42; "Pleads to Maintain Cavalry Strength," *New York Times*, April 17, 1921, 35.

100. "Julian R. Lindsey 1892," accessed October 20, 2014, http://apps.westpointaog.org/Memorials/Article/3481/.

101. J. R. Lindsey to George S. Patton, October 23, 1930, Distribution of Army by Arms of Service 323.2, Box [hereafter B] 10, General Correspondence, 1923–1942, Entry 39 [hereafter E39], Records of the Office of the Chief of Cavalry, Records of the Chiefs of Arms, Record Group [hereafter RG] 177, NACP.

102. General Guy Henry to George S. Patton, October 29, 1930, Distribution of Army by Arms of Service 323.2, B10, E39, RG 177, NACP.

103. Guy V. Henry to Adjutant General, November 26, 1930, Distribution of Army by Arms of Service 323.2, B10, E39, RG 177, NACP. See also Henry to Frank Parker, December 4, 1930, Distribution of Army by Arms of Service 323.2, B10, E39, RG 177, NACP.

104. Guy V. Henry to Fred W. Sladen, November 14, 1930, Distribution of Army by Arms of Service 323.2, B10, E39, RG 177, NACP; Henry to Frank Parker, November 14, 1930, Distribution of Army by Arms of Service 323.2, B10, E39, RG 177, NACP; Henry to Frank McCoy, November 14, 1930, Distribution of Army by Arms of Service 323.2, B10, E39, RG 177, NACP; Henry to Adjutant General, November 26, 1930, Distribution of Army by Arms of Service 323.2, B10, E39, RG 177, NACP.

105. Guy V. Henry to Frank Parker, December 4, 1930, Distribution of Army by Arms of Service 323.2, B10, E39, RG 177, NACP.

106. G. Williams to Adjutant General, July 31, 1934, Distribution of Army by Arms of Service 323.2, B10, E39, RG 177, NACP.

107. H. B. Fiske to Adjutant General, August 13, 1934, Distribution of Army by Arms of Service 323.2, B10, E39, RG 177, NACP.

108. G. Williams to Leon B. Kromer, August 2, 1934, Distribution of Army by Arms of Service 323.2, B10, E39, RG 177, NACP.

109. Leon B. Kromer to George Williams, August 27, 1934, Distribution of Army by Arms of Service 323.2, B10, E39, RG 177, NACP.

CHAPTER 5

1. Odom, *After the Trenches*, 7.

2. *FSR 1923*, 32.

3. Ibid., 36, 34.

4. *Tactics and Technique of Cavalry* (Harrisburg, Pa.: Telegraph Press, 1937), 403.

5. George T. Barnes, "The Necessity for Coordination of Reconnaissance Activities of a Cavalry Division—Aviation—Armored Car" (Combined Arms Research Library, CGSS Student Papers, 1936), 4. See also Ibid., 11; Major Pierce, "The Place of a Mechanized Force in Our Cavalry Organization" (Combined Arms Research Library, CGSS Student Papers, 1932), 18; Doyle O. Hickey, Otto R. Stillinger, and Albert G. Kelly, "Revision of Cavalry, Infantry, and Field Artillery School Pamphlets to Conform to the Present Attack, Defense, and March Reconnaissance Pamphlets" (Combined Arms Research Library, CGSS Student Papers, 1936), 27; James T. Curry Jr., "The Strength, Composition and Missions, and Sphere of Action of: Armored Car Reconnaissance Detachments, Horse Cavalry Reconnaissance Detachments" (Combined Arms Research Library, CGSS Student Papers, 1936), 25.

6. U.S. Army Command and General Staff School, *Reconnaissance Security Marches Halts (Tentative)* (Fort Leavenworth, Kans.: Command and General Staff School Press, 1937), 6.

7. *Other Arms Air Corps* (Fort Riley, Kans.: Academic Division, The Cavalry School, 1931–1932).

8. E. C. McGuire, "Armored Cars in the Cavalry Maneuvers," *CJ* (US) *39* (July 1930): 398.

9. *Reconnaissance Security Marches Halts*, 10–11.

10. Ibid., 10.

11. Terry de la Mesa Allen, *Reconnaissance by Horse Cavalry Regiments and Smaller Units* (Harrisburg, Pa.: Military Service Publishing, 1939), 15.

12. Robert T. Finney, *History of the Air Corps Tactical School, 1920–1940* (Washington, D.C.: Air Force History and Museums Program, 1998), v.

13. J. C. Mullenix, "Form I, Cavalry Course, 1938–1939 School Year," Air Corps Tactical School, 248.80018, AFHRA.

14. Two of the most frequently referenced books in the cavalry course were Command and General Staff School's *Tables of Organization* (Fort Leavenworth, Kans.: Command and General Staff School Press, 1937) and Command and General Staff School's *Tactical Employment of Cavalry* (Fort Leavenworth, Kans.: Command and General Staff School Press, 1935).

15. R. L. Creed, "Tactics and Technique of Cavalry Reconnaissance," lecture, Air Corps Tactical School, November 25, 1936, 248.80017–5, AFHRA.

16. R. L Creed, "Characteristics of Cavalry," lecture, Air Corps Tactical School, December 6, 1937, 248.80017–3, AFHRA.

17. J. C. Mullenix, "The Characteristics and Role of Cavalry," lecture, Air Corps Tactical School, November 3, 1938, 248.80018–2, AFHRA.

18. Hamilton S. Hawkins, "Cavalry and Mechanized Force," *CJ* (US) 40 (September–October 1931): 24; "General Hawkins' Notes," *CJ* (US) 46 (May–June 1937): 239.

19. Note following "Cavalry: A Requisite," *CJ* (US) 47 (May–June 1938): 212.

20. Mullenix, "Characteristics and Role of Cavalry."

21. WO, *FSR Volume I: Organization and Administration, 1930* (London: HMSO, 1930), 1–3.

22. WO, *Cavalry Training Volume I: Training, 1931* (London: HMSO, 1931), 23.

23. Ibid.

24. WO, *FSR Volume II: Operations—General, 1935* (London: HMSO, 1935), 6. (Hereafter *FSR II-Operations*.)

25. *FSR II-Operations*, 62–63.

26. WO, *FSR Volume III: Operations—Higher Formations, 1935* (London: HMSO, 1935), 16.

27. *FSR II-Operations*, 131, 16.

28. John B. Coulter, "Cavalry-Infantry Maneuvers, 1930," *CJ* (US) 39 (July 1930): 349–50; Cuyler L. Clark, "Maneuvers of 1st Cavalry Division," *CJ* (US) 40 (September–October 1931): 10–13.

29. Clark, "Maneuvers of 1st Cavalry Division," 10–13. See also William J. Taylor, "The 305th Cavalry Command Post Exercise," *CJ* (US) 44 (May–June 1935): 24–25.

30. "Army Training Memorandum No. 2 (Collective Training Period, 1930 Supplementary)" (London: WO, 1931), April 8, 1931, 5, WO 231/218, TNA; "Army Training Memorandum, No. 4: Collective Training Period, 1931" (London: HMSO, 1931), December 21, 1931, 10, WO 231/220, TNA; "Army Training Memorandum No. 7: Collective Training Period, 1932" (London: HMSO, 1932), December 13, 1932, 8, WO 231/223, TNA.

31. "Army Training Memorandum No. 10: Collective Training Period, 1933," (London: WO, 1933), December 12, 1933, 7–9, WO 231/226, TNA.

32. "Army Training Memorandum No. 14: Collective Training Period, 1934 (Supplementary) and Individual Training Period, 1934–1935," (London: WO, 1935), May 22, 1935, 20, WO 231/230, TNA.

33. George S. Patton Jr., "The 1929 Cavalry Division Maneuvers," *CJ* (US) 39 (January 1929): 13.

34. McGuire, "Armored Cars," 398.

35. Kinzie Edmunds and Rufus S. Ramey, "6th Cavalry at the Maneuvers of the 8th Brigade," *CJ* (US) 44 (September–October 1935): 29.

36. I. B. Holley, *Army Air Forces Historical Studies: No. 44—Evolution of the Liaison-Type Airplane, 1917–1944* (Washington, D.C.: AAF Historical Office, 1946), 31–32.

37. Edmunds and Ramey, "6th Cavalry at the Maneuvers," 29.

38. Smith, *British Air Strategy*, 4–5.

39. Hinton, *Air Victory*, 378.

40. Roger Douglas Connor, "Grasshoppers and Jump-Takeoff: The Autogiro Programs of the U.S. Army Air Corps" (paper, Sixty-Second Annual Forum and Technology Display of the American Helicopter Society, Phoenix, Ariz., May 9–11, 2006).

41. "Army Training Memorandum No. 2," 19–20.

42. B. Q. Jones, Memorandum for the Commandant Army War College, "Corps and Army Organization-Aviation Components," December 9, 1936, Autogiros 452.1, B50, Document File, 1923–1942, 451.8–454, Records of the Chief of Arms Office-Office of Chief of Cavalry, Records of the Chiefs of Arms, RG 177, NACP.

43. Connor, "Grasshoppers and Jump-Takeoff."

44. Chief of Cavalry Major General Leon B. Kromer to Major Richard L. Creed, December 21, 1937, Folder Equitation-School and Post 1937–1940, 248.80017-9, AFHRA.

45. Adna R. Chaffee, "The Army of the United States," *CJ* (US) 47 (March–April 1938): 110.

46. Ibid., 111.

47. "General Hawkins' Notes: Some Lessons from the War in Spain," *CJ* (US) 47 (March–April 1938): 173.

48. Kinzie B. Edmunds, "The Cavalry-Artillery-Aviation Team," *CJ* (US) 42 (March–April 1933): 11.

49. H. A. Flint, "What a Cavalryman Should Know of the Air," *CJ* (US) 42 (July–August 1933): 9.

50. Letter to Adjutant General from Colonel A. M. Miller Jr. for Chief of Cavalry, December 12, 1936, Autogiros 452.1, B50, E39, RG 177, NACP. See also Letter to Adjutant General from Colonel A. M. Miller Jr., April 10, 1937, Autogiros 452.1, B50, E39, RG 177, NACP.

51. H. F. Gregory, *Anything a Horse Can Do: The Story of the Helicopter* (New York: Reynal and Hitchcock, 1944), 53.

52. Gregory, *Anything a Horse Can Do*, 50–51. For an additional description of the way autogiros worked with diagrams, see L. J. McNair, "And Now the Autogiro," *Field Artillery Journal* 28 (1937): 3–17.

53. Bruce H. Charnov, *From Autogiro to Gyroplane: The Amazing Survival of an Aviation Technology* (Westport, Conn.: Praeger, 2003), 101–2.

54. Ibid., 95.

55. Juan de la Cierva and Don Rose, *Wings of Tomorrow: The Story of the Autogiro* (New York: Brewer, Warren and Putnam, 1931), 49–50.

56. "Army Training Memorandum No. 11: Collective Training Period, 1933 (Supplementary)," May 18, 1934, 18, WO 231/227, TNA.

57. Special cable to the *Inquirer*, "British to Use 'Gyros,'" December 14, 1934, Autogiros 452.1, B50, E39, RG 177, NACP.

58. Charnov, *From Autogiro to Gyroplane*, 131, 189.

59. Ibid., 189. British aircraft designations do not use letters or numbers as the primary identifier for aircraft. Thus, Avro Rota identifies a machine called the Rota that was produced by Avro. The Avro Rota was a British version of the Spanish Cierva C.30A, manufactured under license.

60. "Army Training Memorandum No. 18, April 1937," March 24, 1937, 26, WO 231/234, TNA.

61. "Army Training Memorandum No. 22: Collective Training Period, 1939," April 18, 1939, 26, WO 231/238, TNA.

62. W. Wallace Kellett, "Autogiro Development in Europe," n.d. [1937?], Autogiros 452.1, B50, E39, RG 177, NACP.

63. Charnov, *From Autogiro to Gyroplane*, 188, 199.

64. Colonel E. R. W. McCabe War Department General Staff Military Intelligence Division to Lieutenant Colonel V. W. B. Wales, 1st Cavalry, Fort Knox, Kentucky, January 10, 1940, Autogiros 452.1, B50, E39, RG 177, NACP.

65. Peter W. Brooks, *Cierva Autogiros: The Development of Rotary-Wing Flight* (Washington, D.C.: Smithsonian Institution Press, 1988), 192, 254.

66. W. David Lewis, "The Autogiro Flies the Mail! Eddie Rickenbacker, Johnny Miller, Eastern Air Lines, and Experimental Airmail Service with Rotorcraft, 1939–1940," in *Realizing the Dream of Flight: Biographical Essays in Honor of the Centennial of Flight, 1903–2003*, ed. Virginia Parker Dawson and Mark D. Bowles (Washington, D.C.: National Aeronautics and Space Administration History Division, Office of External Relations, 2005): 75–77, 80.

67. Adjutant General Wm. F. Pearson to Chief of the Air Corps, Subject-Autogiros, June 11, 1934, Autogiros 452.1, B50, E39, RG 177, NACP.

68. Lewis, "Eddie Rickenbacker," 80.

69. James G. Ray, "Military and Naval Uses of Autogiros," n.d. [1935?], Autogiros 452.1, B50, Document File, 1923–1942, 451.8–454, Office of Chief of Cavalry, Records of the Chiefs of Arms, RG 177, NACP.

70. Connor, "Grasshoppers and Jump-Takeoff."

71. *Hearings on H.R. 8143: To Authorize the Appropriation of Funds for the Development of the Autogiro, Before the House Committee on Military Affairs*, Seventy-Fifth Cong. 88 (1938).

72. Cierva and Rose, *Wings of Tomorrow*, 49–50.

73. Connor, "Grasshoppers and Jump-Takeoff."

74. Holley, *Army Air Forces Historical Studies,* 45.

75. Letter to Adjutant General from Executive Cavalry Colonel A. M. Miller Jr., "Subject: Kellett Autogiro," July 29, 1935, Autogiros 452.1, B50, E39, RG 177, NACP.

76. Letter to Chief of the Air Corps from Leon B. Kromer, Office Chief of Cavalry, War Department, August 9, 1935, Autogiros 452.1, B50, E39, RG 177, NACP.

77. Letter to the Adjutant General from Colonel Bruce Palmer, September 19, 1936, Autogiros 452.1, B50, E39, RG 177, NACP.

78. Letter to Adjutant General from Colonel A. M. Miller Jr., Office Chief of Cavalry, October 9, 1936, Autogiros 452.1, B50, E39, RG 177, NACP.

79. Letter to Adjutant General from Colonel A. M. Miller Jr. for Chief of Cavalry, December 12, 1936. See also Letter to Adjutant General from Colonel A. M. Miller Jr., April 10, 1937.

80. To Chief of Cavalry from Brigadier General Guy V. Henry, March 13, 1937, Autogiros 452.1, B50, E39, RG 177, NACP.

81. Charnov, *From Autogiro to Gyroplane,* 133.

82. Letter to the Adjutant General from Lieutenant Colonel H. C. Davidson, Air Corps Executive, "Subject: Schedule for Autogiros," June 25, 1937, Autogiros 452.1, B50, E39, RG 177, NACP.

83. Gregory, *Anything a Horse Can Do,* 79.

84. Guy V. Henry, "Report of Test of Autogiro," June 10, 1937, Autogiros 452.1, B50, E39, RG 177, NACP.

85. Cavalry Board to Chief of Cavalry, "Report of Test of Autogiro," June 10, 1937, Autogiros 452.1, B50, E39, RG 177, NACP.

86. Captain E. T. Rundquist, "Report of Commanding Officer, Det. First Observation Squadron, Air Corps, on Scope of Tests Conducted and the Suitability of the Autogiro for Military Use as Compared with the Airplane," June 16, 1937, Autogiros 452.1, B50, E39, RG 177, NACP; Guy V. Henry to Chief of Cavalry, "Report of Test of Autogiro by the Cavalry Board, Fort Riley, Kansas," June 10, 1937, Autogiros 452.1, B50, E39, RG 177, NACP.

87. *Hearings on H.R. 8143,* 56.

88. Ibid., 55.

89. Frank Dorsey, 76th Congress 1st Session-H. R. 8143 "A Bill-to Authorize the Appropriation of Funds for the Development of the Autogiro," August 4, 1937, Autogiros 452.1, B50, E39, RG 177, NACP.

90. Dorsey R. Rodney, "Report on the Progress of Test of Autogiros" to Commandant, the Cavalry School, Feb 8, 1939, Autogiros 452.1, B50, E39, RG 177, NACP.

91. Ibid., 56.

92. War Department, Office Chief of Cavalry, Washington letter to the Adjutant General, August 8, 1938, Autogiros 452.1, B50, E39, RG 177, NACP.

93. Colonel Charles Burnett to R. G. Kellett, Kellett Autogiro Corporation, August 21, 1939, Autogiros 452.1, B50, E39, RG 177, NACP.

94. C. H. Unger, Special Orders No. 66, Proceedings of a Board of Officers, Headquarters Seventh Cavalry Brigade, Fort Knox, Kentucky, October 10, 1939, Autogiros 452.1, B50, E39, RG 177, NACP.

95. Robert M. Lee to Chief of the Material Division, A. C. Wright Field, Dayton, Ohio, "Subject: Autogyro Data," November 7, 1939, Autogiros 452.1, B50, E39, RG 177, NACP.

96. Colonel Jack W. Heard, Board meeting minutes, February 20, 1940, Autogiros 452.1, B50, E39, RG 177, NACP.

97. Letter to the Chief of Cavalry from Brigadier General Robert C. Richardson Jr., March 22, 1940, Autogiros 452.1, B50, E39, RG 177, NACP.

98. M. F. Davis to Assistant Chief of Staff, G-3; Assistant Chief of Staff, G-4; Chief of Field Artillery; Chief of Cavalry; and Chief of Infantry, in turn, "Subject: Proposed Corps and Army Observation Airplane," December 28, 1937, Autogiros 452.1, B50, E39, RG 177, NACP; Colonel A. M. Miller Jr. for the Chief of Cavalry to the Chief of Infantry, February 2, 1938, Autogiros 452.1, B50, E39, RG 177, NACP.

99. The Crouch-Bolas B-42 should not be confused with the Douglas XB-42, which appeared late in World War II.

100. William D. Crittenberger to Major Topkins, June 25, 1940, Autogiros 452.1, B50, E39, RG 177, NACP.

101. For detailed descriptions of the development of mechanization in the United States and Great Britain, see Alexander Magnus Bielakowski, "U.S. Army Officers and the Issue of Mechanization, 1920–1942" (PhD diss., Kansas State University, 2002); Robert S. Cameron, "Americanizing the Tank: U.S. Army Administration and Mechanized Development within the Army, 1917–1943" (PhD diss., Temple University, 1994); Mildred Hanson Gillie, *Forging the Thunderbolt: A History of the Development of the Armored Force* (Harrisburg, Pa.: Military Service Publishing, 1947); Barton C. Hacker, "The Military and the Machine: An Analysis of the Controversy over Mechanization in the British Army, 1919–1939" (PhD diss., University of Chicago, 1968); Basil H. Liddell Hart, *The Tanks: The History of the Royal Tank Regiment and Its Predecessors, Heavy Branch Machine-Gun Corps, Tank Corps, and Royal Tank Corps*, vol. 1, *1914–1945* (London: Cassell, 1959); Hofmann, *Through Mobility We Conquer*; Johnson, *Fast Tanks and Heavy Bombers*; Vincent J. Tedesco III, "'Greasy Automatons' and the 'Horsey Set': The U.S. Cavalry and Mechanization, 1928–1940." (master's thesis, Pennsylvania State University, 1995).

102. George S. Patton Jr. and C. C. Benson, "Mechanization and Cavalry," *CJ* (US) 39 (April 1930): 234–40.

103. K. B. Edmunds, "Tactics of a Mechanized Force: A Prophecy," *CJ* (US) 39 (July 1930): 414.

104. M. N. MacLeod, "This Tank Business—in Fact and Fancy," *CJ* (US) 42 (January–February 1933): 48.

105. George S. Patton Jr. "Motorization and Mechanization in the Cavalry," *CJ* (US) 39 (July 1930): 335.

106. Ibid., 336.

107. Patton, "Motorization and Mechanization," 334.

108. Ibid., 331–33.

109. Charles L. Scott, "Are More Changes Needed in Our Horsed Cavalry Regiment Now?" *CJ* (US) 44 (September–October 1935): 42.

110. "Progress and Discussion," *CJ* (US) 39 (January 1930): 118.

111. Scott, "Are More Changes Needed?" 42.

112. Taylor, "305th Cavalry Command Post Exercise," 24.

113. Patton, "1929 Cavalry Division Maneuvers," 9.

114. "A. F. V.," "Cavalry and Tanks," *CJ* (UK) 24 (1934): 454. See also *Reconnaissance Security Marches Halts*, 11. See also Adna Chaffee, "Mechanization in the Army," lecture, The Army War College, Washington, D.C., September 29, 1939, Combined Arms Library Digital Library, 5, 9; "Iron Horses for the Cavalry," *Literary Digest* (March 19, 1932), 41; Thomas J. Johnson, "A Horse! A Horse! My Kingdom for a Horse," *Quartermaster Review* 16 (September–October 1936): 6; R. L. Creed, "Tactics and Technique of Cavalry Reconnaissance," lecture, Air Corps Tactical School, November 25, 1936, 248.80017-5, AFHRA.

115. "Cavalry: A Requisite," 212.

116. "Machines, Maneuvers, and Mud," reprint from editorial in *The Milwaukee Journal* in *CJ* (US) 48 (January–February 1939): 73.

117. Chaffee, "Mechanization in the Army," 7.

118. William R. Irvin, *Combined Reconnaissance Operation by Cavalry and Aviation* (Fort Leavenworth, Kans.: Command and General Staff School, 1934), 5. See also *Tactics and Technique of Cavalry: A Text and Reference Book of Cavalry Training (Includes All Technical Changes to June, 1937)*, 8th ed. (Harrisburg, Pa.: Military Service Publishing, 1937), 256.

119. Reginald Hargreaves, "Crisis in the Cavalry," *CJ* (UK) 28 (1938): 596.

120. Henry Cabot Lodge Jr., "Cavalry 'Marches' on Wheels: A Description of the Big Bend Portée," *Army Ordnance* 14 (November–December 1933): 135.

121. "Machines, Maneuvers, and Mud," 73. Emphasis added.

122. John B. Smith, "The Effect of Mechanization Upon Cavalry," *CJ* (US) 40 (November–December 1931): 21, 24. See also "Foreign Magazines," *CJ* (UK) 22 (1932): 298–99.

123. Bernard Lentz, "A Justification of Cavalry," *CJ* (US) 44 (January–February 1935): 10.

124. "Foreign Views on Mechanization," *CJ* (US) 40 (July–August 1931): 64.

125. Chaffee, "Mechanization in the Army," 5.

126. "Army Training Memorandum No. 4," 31.

127. Pierce, "Place of a Mechanized Force," 18.

128. Irvin, *Combined Reconnaissance Operation*, 4.

129. Edmunds and Ramey, "6th Cavalry at the Maneuvers," 28.

130. *Tactics and Technique of Cavalry: A Text*, 9, 14.

131. Ibid., 256.

132. Hamilton S. Hawkins, "Some Observations on the Attack by Combined Arms," *CJ* (US) 47 (March–April 1938): 147.

133. "General Hawkins' Notes: The Combination of Horse Cavalry with Mechanized Cavalry," *CJ* (US) 47 (September–October 1938): 462.

134. "Cavalry Affairs before Congress," *CJ* (US) 48 (March–April 1939): 130.

135. Ibid., 130–131.

136. Ibid., 133, 135.

137. "Exit the Cavalry . . . Enter the Tanks," *Popular Science Monthly* 119 (August 1931), 40–41.

138. "Army Training Memorandum No. 6: Individual Training Period, 1931–32," May 6, 1932, 8, WO 231/222, TNA.

139. Edmunds and Ramey, "6th Cavalry at the Maneuvers," 28.

140. "Army Training Memorandum No. 17: Collective Training Period, 1936," January 7, 1937, 18, WO 231/233, TNA.

141. "Horses and Motors," reprint from the (Springfield) *Illinois State Journal* in *CJ* (US) 45 (March–April 1936): 105.

142. Ibid.

143. "Army Training Memorandum No. 5: Collective Training Period, 1931 (Supplementary)" (London: HMSO, 1932), April 21, 1932, 7, WO 231/221, TNA.

144. "Report by a Committee Assembled to Consider the Organization of the Mechanized Cavalry and the Royal Tank Corps," n.d. [1938?], WO 33/1509, TNA.

145. E. G. Hume, "Cavalry Today—Horse v. Machine?" *CJ* (UK) 20 (1930): 180.

146. "Machines, Maneuvers, and Mud," 73.

147. "Value of Animals in Modern Warfare," *CJ* (US) 47 (March–April 1938): 109.

148. Hume, "Cavalry Today," 180.

149. R. W. Grow, "One Cavalry," *CJ* (US) 47 (March–April 1938): 150.

150. "A. F. V." "Cavalry and Tanks," 449.

151. "The Editor's Saddle: Major General Leon B. Kromer," *CJ* (US) 47 (March–April 1938): 164.

152. Ibid.

153. "Characteristics of Cavalry," Lecture, Conference No. 4, December 6, 1937, Course: Cavalry, The Air Corps Tactical School, Maxwell Field, Alabama, 1937–1938, Folder Cavalry Form II, 1937–1938, No. 4 248.80017-3, 1937–1938, AFHRA. Emphasis in original.

154. John K. Herr, "What of the Future?" *CJ* (US) 48 (January–February 1939), 3.

155. John K. Herr, "My Greetings to All Cavalrymen," *CJ* (US) 47 (March–April 1938), 99.

156. Herr, "What of the Future?" 4.

157. "Cavalry Affairs before Congress," 130.

158. *Tactics and Technique of Cavalry: A Text,* 14.

159. Katzenbach, "Horse Cavalry in the Twentieth Century," 138.

160. Hansard HC Deb 08 March 1932, vol. 262, col. 1683.

161. Ibid.

162. Ibid., 1685.

163. Hansard HC Deb 19 March 1936, vol. 310, col. 729.

164. Ibid.

165. Ibid., 730.

166. Hansard HC Deb 12 March 1936, vol. 309, col. 2457.

167. Ibid., 2466.

168. Ibid., 2467.

169. Hansard HC Deb 19 March 1936, vol. 301, cols. 732–33.

170. Linn, *The Echo of Battle,* 6.

171. Ibid., 66–67.

172. Ibid., 6–7.

173. "Night Guard," *CJ* (UK) 28 (1938): 170–71.

174. Cyril Stacey, "The Passing of the Cavalry, 1939," *CJ* (UK) 29 (January–October 1939): 566.

175. Katzenbach, *Horse Cavalry,* 140.

176. Mark Hampton, "Inventing David Low: Self-Presentation, Caricature and the Culture of Journalism in Mid-Twentieth Century Britain," *Twentieth Century British History* 20, no. 4 (2009): 483. This Colonel Blimp should not be confused with the title character of the 1943 British film *The Life and Death of Colonel Blimp,* although the name was taken from Low's popular cartoon creation.

177. Peter Mellini, "Colonel Blimp's England," *History Today* 34 (October 1984): 10, http://www.powell-pressburger.org/Reviews/43_Blimp/Blimp18.html.

178. Colin Seymour-Ure and Jim Schoff, *David Low* (London: Secker and Warburg, 1985), 136.

179. Ibid. See also Colin Fleming, "The Greatest British Film Ever Is 'The Life and Death of Colonel Blimp'," *The Atlantic,* March 27, 2013, http://www.theatlantic.com /entertainment/archive/2013/03/the-greatest-british-film-ever-is-the-life-and-death-of -colonel-blimp/274381/.

180. Seymour-Ure and Schoff, *David Low,* 94.

181. Ibid., 94, 133.

182. C. S. Forester, *The General* (1936; repr., Annapolis, Md.: Nautical and Aviation Publishing, 1982), x.

183. Ibid., 24.

184. Ibid., 25.

185. Ibid.

186. Bielakowski, "U.S. Army Officers," 93.

187. Hawkins, "Cavalry and Mechanized Force," 25.

188. "Editorial," *CJ* (UK) 28 (April 1938): 163.

189. Ibid., 164.

CONCLUSION

1. Bielakowski, "U.S. Army Officers," 93.

2. John K. Herr and Edward S. Wallace, *The Story of the U.S. Cavalry, 1775–1942,* (Boston: Little, Brown, 1953): 261.

3. Mark Anderson and Greg Barker, "Campaign Against Terror," *Frontline,* September 8, 2002, http://www.pbs.org/wgbh/pages/frontline/shows/campaign/ground/warstories. html.

4. Stephen Biddle, *Afghanistan and the Future of Warfare: Implications for Army and Defense Policy* (Carlisle, Pa.: Strategic Studies Institute, 2002): 9–10. See also Doug Stanton, *Horse Soldiers: The Extraordinary Story of a Band of U.S. Soldiers Who Rode to Victory in Afghanistan* (New York: Scribner, 2009).

5. "Attack of the Drones," *Newsweek,* September 19, 2009, http://www.newsweek .com/2009/09/18/attack-of-the-drones.html.

6. As quoted in "The Death of the F-22 Fighter Plane," *Newsweek,* September 18, 2009, http://www.newsweek.com/death-f-22-fighter-plane-79281.

BIBLIOGRAPHY

ARCHIVE SOURCES

Air Force Historical Research Agency, Maxwell Air Force Base, Alabama
    Air Corps Tactical School Collection
    Army Air Service Collection
    Benjamin D. Foulois Papers
    Frank Purdy Lahm Collection
    Grover O'Neill Collection

The British National Archives, London
    Air Ministry and Royal Air Force Records
    Cabinet Office Series
    Public Record Office Series
    Treasury
    War Office Series

Military History Institute, Carlisle, Pa.
    United States Army, Air Service. *Annual Report of the Chief of Air Service for the Fiscal Year ending June 30, 1922.* Washington, D.C.: GPO, 1922.
    Paul W. Beck, Lecture on Aeroplanes, United States Army.
    American Expeditionary Forces. *Report of Superior Board on Organization and Tactics.* United States?: s.n., 1919.
    Cavalry School Department of Tactics. *Employment of Cavalry, 1924–1925.* Fort Riley, Kans.: Cavalry School, Dept. of Tactics, 1925.
    World War I Veterans Survey Collection, 1898–1941

National Archives 2, College Park, Maryland
    Office of the Chief of Cavalry, Records of the Chiefs of Arms, RG 177

PUBLISHED PRIMARY SOURCES

"The Action of the Future." *Journal of the United States Cavalry Association* 19 (January 1909): 635–41.

"The Aeroplane in War." *Journal of the United States Cavalry Association* 21 (November 1910): 536–39.

"The Aeroplanes and the Cavalry." *Cavalry Journal* (US) 23 (July 1912): 123–25.

"A. F. V." "Cavalry and Tanks." *Cavalry Journal* (UK) 24 (1934): 449–55.

Air Ministry-Air Historical Branch (Great Britain). *A Short History of the Royal Air Force.* [S.I.] Air Ministry, 1936.

Allen, Terry de la Mesa. *Reconnaissance by Horse Cavalry Regiments and Smaller Units.* Harrisburg, Pa.: Military Service Publishing, 1939.

Allenby, E. H. H. "Cavalry in Future Wars." *Cavalry Journal* (US) 30 (January 1921): 1–2.

"And Now the Autogiro." *Field Artillery Journal* 28 (1937): 3–17.

Anderson, Edward. "Comment and Criticism on 'Mounted Rifles.'" *Journal of the United States Cavalry Association* 13 (April 1903): 719–22.

Andrus, E. P. "The Saber." *Journal of the United States Cavalry Association* 4 (December 1892): 373–82.

Annual Report of the Commandant General Service and Staff College for the School Year Ending August 31, 1904. Fort Leavenworth, Kans.: Staff College Press, 1904.

Annual Report of the Commandant Infantry and Cavalry School and Staff College for the School Year Ending August 31, 1905. Fort Leavenworth, Kans.: Staff College Press, 1905.

Annual Report of the Commandant U.S. Infantry and Cavalry School, U.S. Signal School and Army Staff College for the School Year Ending August 31, 1907. Fort Leavenworth, Kans.: Staff College Press, 1907.

Arnold, H. H. "The Cavalry-Air Corps Team." *Cavalry Journal* (US) 37 (January 1928): 70–76.

Augur, J. A., C. B. Hoppin, E. A. Godwin, S. D. Rockenbach, M. C. Butler, Eugene A. Carr, John Bigelow Jr., et. al. "Comments on Mounted Rifles." *Journal of the United States Cavalry Association* 13 (January 1903): 386–407.

Augur, J. A., Edward Anderson, and Cornelius C. Smith, "Comment and Criticism-Mounted Rifles." *Journal of the United States Cavalry Association* 13 (April 1903): 718–23.

Bailes, K. E. "Technology and Legitimacy: Soviet Aviation and Stalinism in the 1930's." *Technology and Culture* 17 (January 1976): 55–81.

Bannerman-Phillips, H. "Air-craft in Co-operation with Cavalry." *Nineteenth Century and After* 69 (1911): 798–810.

———. "The Future of Airships in War." *United Service Magazine* 37 (September 1908): 589–94.

Barnum, Malvern-Hill. "A Chief of Cavalry." *Journal of the United States Cavalry Association* 15 (April 1905): 946–49.

Barrow, George. "The Future of Cavalry." *Cavalry Journal* (UK) 19 (January–October 1929): 176–84.

Barrows, John Stuart. "The Uhlans and Other Cavalry in the European War." *Journal of the United States Cavalry Association* 26 (January 1916): 390–98.

Battine, Cecil. "The German Cavalry Manoevures of 1911." *Cavalry Journal* (UK) 7 (January 1912): 47–52.

Bell, M. S. E. Harry. "The First Aeroplane under Rifle Fire." *Journal of the United States Cavalry Association* 22 (May 1912): 1128–30.

Benson, C. C. "Mechanization Aloft and Alow." *Cavalry Journal* (US) 38 (January 1929): 58–62.

Bernhardi, Friedrich von. *Cavalry in Future Wars.* London: John Murray, 1909.

———. *Cavalry in War and Peace.* London: H. Rees, 1910.

Bigelow, John, Jr. *The Campaign of Chancellorsville, a Strategic and Tactical Stud.* New Haven, Conn.: Yale University Press, 1910.

———. *The Principles of Strategy: Illustrated Mainly from American Campaigns.* 2nd ed. Philadelphia: J. B. Lippincott, 1894.

———. *Reminiscences of the Santiago Campaign.* New York: Harper and Brothers, 1899.

———. "The Sabre and Bayonet Question." *Journal of the Military Service Institute of the United States* 3 (1882): 65–96.

Blunt, J. Y. "The Shock Action of Cavalry." *Journal of the United States Cavalry Association* 5 (1892): 33–45.

"Boots and Saddles [pseud.]." "Cavalry Equipment." *Journal of the United States Cavalry Association* 21 (March 1911): 966–69.

Borton, A. E. "The Use of Aircraft in Small Wars." *Journal of the Royal United Service Institution* 65 (1921): 310–19.

Bradford, K. S. "Modern Cavalry." *Infantry Journal* 35 (December 1929): 558–66.

Brechard, General. "The French Cavalry." *Cavalry Journal* (US) 38 (October 1929): 486–517.

Brown, H. Clifton. "The Training of Cavalry in Battle." *Cavalry Journal* (UK) 7 (January–October 1912): 156–60.

Butler, M. C. "The Saber." *Journal of the United States Cavalry Association* 14 (July 1903): 142–44.

Cammell, R. A. "Aeroplanes with Cavalry." *Cavalry Journal* (UK) 6 (April 1911): 197–99.

Campbell, R. A. "Aeroplanes with Cavalry." *Cavalry Journal* (US) 22 (September 1911): 311–14.

Carter, William Harding. *Horses, Saddles and Bridles.* Leavenworth, Kans.: Ketcheson and Reeves, 1895.

"Cavalry Affairs before Congress." *Cavalry Journal* (US) 48 (March–April 1939): 130–135.

"Cavalry and the Mounted Infantry." *Journal of the United States Cavalry Association* 17 (October 1906): 363–66.

"Cavalry, an Essential Arm." *Cavalry Journal* (UK) 12 (January 1922): 6–8.

"Cavalry and the Aeroplane." *Journal of the United States Cavalry Association* 20 (November 1909): 617.

"'Cavalry Journal' Committee." *Cavalry Journal* (UK) 16 (1926): 1–3.

A Cavalry Officer Abroad. "Report Upon Year Spent with French Cavalry." *Journal of the United States Cavalry Association* 25 (January 1915): 446–47.

———. "The Question of Organization." *Journal of the United States Cavalry Association* 25 (October 1914): 199–220.

Cavalry School (U.S.) Academic Division. *Cavalry Operations during the World War.* Fort Riley, Kans.: Academic Division, The Cavalry School, 1929.

"The Cavalry Spirit." *Cavalry Journal* (UK) 12 (January–October 1922): 225.

Chaffee, Adna R. "The Army of the United States." *Cavalry Journal* (US) 47 (March–April 1938): 110–13.

———. "The Maneuvers of the First Cavalry Division, September–October 1923." *Cavalry Journal* (US) 33 (April 1924): 133–62.

Charrington, H. V. S. "Where Cavalry Stands To-day." *Cavalry Journal* (UK) 17 (January–October 1927): 419–30.

Childers, Erskine. *German Influence on British Cavalry.* London: E. Arnold, 1911.

———. *War and the Arme Blanche.* London: E. Arnold, 1910.

Cierva, Juan de la, and Don Rose. *Wings of Tomorrow: The Story of the Autogiro.* New York: Brewer, Warren and Putnam, 1931.

Clark, Cuyler L. "Maneuvers of 1st Cavalry Division." *Cavalry Journal* (US) 40 (September–October 1931): 10–18.

Colby, Elbridge. "The Horse in War Today and Yesterday: I—Cavalry Still a Factor in Military Warfare." *Current History* 28 (1928): 446–52.

"Comment and Criticism-Mounted Rifles." *Journal of the United States Cavalry Association* 13 (April 1903): 718–23.

"Constitution of the United States Cavalry Association." *Journal of the United States Cavalry Association* 13 (July 1902): 97–100.

Contributed. "Cavalry Organization." *Cavalry Journal* (US) 29 (July 1920): 116–21.

Cooke, Philip St. George. *Cavalry Tactics or Regulations for the Instruction, Formation, and Movements of the Cavalry of the Army and Volunteers of the United States.* Washington, D.C.: Government Printing Office, 1861–62.

———. "Our Cavalry." *United Service* 3 (July 1879): 329–346.

Coulter, John B. "Cavalry-Infantry Maneuvers, 1930." *Cavalry Journal* (US) 39 (July 1930): 349–65.

Croft, W. D. "Notes on Armoured Cars." *Cavalry Journal* (UK) 15 (1925): 160–75.

Crombie, D. C. "Cavalry in Frontier Warfare." *Cavalry Journal* (UK) 5 (1910): 358–69.

Crosby, Herbert B. "Cavalry of Today." *Cavalry Journal* (US) 36 (October 1927): Frontispiece.

Cruikshank, J. R. H. "How the Cavalry Exploits a Victory: Being Extracts from the Diary of a Subaltern under Allenby in Palestine." *Cavalry Journal* (US) 32 (April 1923): 163–70.

Dade, A. L. "Reducing the Cavalry of the Regular Army." *Journal of the United States Cavalry Association* 22 (May 1912): 1010–16.

Davis, Edward. "The British Cavalry in Palestine and Syria." Pt. 1. *Cavalry Journal* (US) 31 (April 1922): 123–29.

———. "The British Cavalry in Palestine and Syria." Pt. 2. *Cavalry Journal* (US) 31 (July 1922): 270–77.

———. "The British Cavalry in Palestine and Syria." Pt. 3. *Cavalry Journal* (US) 31 (October 1922): 395–402.

———. "The British Cavalry in Palestine and Syria." Pt. 4. *Cavalry Journal* (US) 31 (January 1923): 56–65.

———. "The British Cavalry in Palestine and Syria: Conquest of Syria (continued)." Pt. 5. *Cavalry Journal* (US) 32 (October 1923): 435–44.

Davis, George B. "The Reorganization of the Cavalry." *Journal of the United States Cavalry Association* 22 (March 1912): 797–805.

Davison, Paul. "The Cavalry's Cavalry: A Study of Armored Cars, Their Powers and Limitations, and Some Notes as to Their Tactical Employment." *Cavalry Journal* (US) 36 (January 1927): 90–95.

De la Vergne, Captain. "A Few Words about European Cavalry." *Cavalry Journal* (US) 29 (July 1920): 152–57.

De Lisle, H. de B. "The Strategical Action of Cavalry." *Cavalry Journal* (UK) 7 (January–October 1912): 320–34.

Denison, George T. *A History of Cavalry from the Earliest Times with Lessons for the Future.* 2nd ed. London: Macmillan, 1913.

Dillman, George. "1st Cavalry Division Maneuvers." *Cavalry Journal* (US) 37 (January 1928): 47–65.

———. "Joint Cavalry and Air Corps Training." *Cavalry Journal* (US) 36 (October 1927): 560–67.

"Dominion and Foreign Magazines." *Cavalry Journal* (UK) 15 (January–October 1925): 96–98.

Donaldson, Thomas Q. "The Revolver or Pistol Best Suited to Cavalry." *Journal of the United States Cavalry Association* 13 (July 1902): 71–74.

Drill Regulations for Cavalry, United States Army: Amended 1909, Corrected to January 1. Washington, D.C.: Government Printing Office, 1911.

Dwan, Edward J. "The Reorganization of the United States Cavalry." *Cavalry Journal* (UK) 19 (January–October 1929): 602–15.

A Dweller on Earth. "A Little Bird Told Me." *Cavalry Journal* (UK) 18 (1928): 601–4.

"Editorial." *Cavalry Journal* (UK) 28 (April 1938): 163–64.

"Editorial and Professional Notes: Military Aeroplanes." *Journal of the United States Artillery* 37 (January–February 1912): 102.

"Editorial Comment: A Cavalry Army." *Cavalry Journal* (US) 30 (October 1921): 418–20.

"Editorial Comment: The Cavalry Journal Reappears." *Cavalry Journal* (US) 29 (April 1920): 81–83.

"Editorial Comment: Modern Defense Policies Increase Importance of Cavalry." *Cavalry Journal* (US) 32 (October 1923): 450–51.

"The Editor's Saddle: Major General Leon B. Kromer." *Cavalry Journal* (US) 47 (March–April 1938): 164–67.

Editor's Table. "Chief of Cavalry." *Journal of the United States Cavalry Association* 19 (July 1908): 197–98.

Edmunds, Kinzie B. "The Cavalry-Artillery-Aviation Team." *Cavalry Journal* (US) 42 (March–April 1933): 7–11.

———. "Tactics of a Mechanized Force: A Prophecy." *Cavalry Journal* (US) 39 (July 1930): 410–17.

Edmunds, Kinzie B., and Rufus S. Ramey. "6th Cavalry at the Maneuvers of the 8th Brigade." *Cavalry Journal* (US) 44 (September–October 1935): 21–29.

Eltinge, LeRoy. "Review of Our Cavalry Situation." *Cavalry Journal* (US) 29 (April 1920): 14–22.

E. S. "Chief of Cavalry." *Journal of the United States Cavalry Association* 18 (January 1908): 555–56.

Fanshane, H. D. "Cavalry in Mesopotamia in 1918." *Cavalry Journal* (UK) 10 (January–October 1920): 414–29.

Farman, Elbert E. "The Cavalry in the Present War." *Journal of the United States Cavalry Association* 26 (April 1916): 625–43.

Fickett, Edward M. "A Study of the Relationship between the Cavalry and the Air Service in Reconnaissance." *Cavalry Journal* (US) 32 (October 1923): 412–22.

Flint, H. A. "What a Cavalryman Should Know of the Air." *Cavalry Journal* (US) 42 (July–August 1933): 5–11.

Foltz, F. S. "The Necessity for Well Organized Cavalry." *Journal of the United States Cavalry Association* 23 (March 1913): 723–45.

Formby, John. *Cavalry in Action in the Wars of the Future.* London: H. Rees, 1905.

"Foreign Magazines." *Cavalry Journal* (UK) 18 (January–October 1928): 664–68.

"Foreign Magazines." *Cavalry Journal* (UK) 22 (January–October 1932): 298–300.

"Foreign Views on Mechanization." *Cavalry Journal* (US) 40 (July–August 1931): 39–40, 64.

Foulois, Benjamin F. "Military Aviation and Aeronautics." *Journal of the Military Service Institution of the United States* 52 (1913): 98–119.

Foulois, Benjamin D., with C. V. Glines. *From the Wright Brothers to the Astronauts: The Memoirs of Major General Benjamin D. Foulois.* New York: McGraw-Hill, 1968.

Fuller, Ezra B., Jr. "More Cavalry Instead of Less." *Journal of the United States Cavalry Association* 22 (March 1912): 972–83.

———. "Why We Need Cavalry." *Journal of the United States Cavalry Association* 22 (March 1912): 966–69.

"Fundamentals of Cavalry Training Policy: Cavalry Memorandum No. I, War Department, December 10, 1920." *Cavalry Journal* (US) 30 (April 1921): 180–85.

"The Future of Cavalry: A Lecture Given at the University of Bristol on 7th February, 1929." *Cavalry Journal* (UK) 19 (January–October 1929): 365–79.

Gaston, J. A. "Divisional Cavalry on the March and in Action." *Journal of the United States Cavalry Association* 25 (April 1915): 600–607.

"General Hawkins' Notes." *Cavalry Journal* (US) 46 (May–June 1937): 239.

"General Hawkins' Notes: The Combination of Horse Cavalry with Mechanized Cavalry." *Cavalry Journal* (US) 47 (September–October 1938): 461–62.

"General Hawkins' Notes: Some Lessons from the War in Spain." *Cavalry Journal* (US) 47 (March–April 1938): 173–74.

Gerlach, William. "Thoughts about Cavalry." *Journal of the United States Cavalry Association* 13 (January 1903): 372–79.

Gilpin, E. H. "Armament and Equipment of the Cavalryman." *Journal of the United States Cavalry Association* 22 (July 1911): 76–84.

Golovine, N. N. "Cavalry on the Flank and in the Rear." *Cavalry Journal* (US) 31 (January 1922): 40–59.

———. "Cavalry Reconnaissance: The Modern Service of Information and Cavalry's Role in It." *Cavalry Journal* (US) 31 (April 1922): 184–94.

———. "Modern Cavalry and Its Organization." *Cavalry Journal* (US) 35 (January 1926): 29–39.

Goltz, Colmar Freiherr von der. *The Conduct of War: A Short Treatise on Its Most Important Branches and Guiding Rules.* Translated by G. F. Leverson. London: K. Paul Trench, Trübner and Co, 1899.

Gray, Alonzo. *Cavalry Tactics: As Illustrated by the War of the Rebellion Together with Many Interesting Facts Important for Cavalry to Know.* Fort Leavenworth, Kans.: U.S. Cavalry Association, 1910.

Griepenkerl, Otto. *Letters on Applied Tactics.* Kansas City, Kans.: Hudson Press, 1906.

Grow, R.W. "One Cavalry." *Cavalry Journal* (US) 47 (March–April 1938): 150.

Haig, Earl. "Introductory Remarks." *Cavalry Journal* (UK) 10 (April 1920): 5–6.

Harbord, James. "Cavalry in Modern War." *Journal of the United States Cavalry Association* 15 (April 1905): 765–71.

Hargreaves, Reginald. "Crisis in the Cavalry." *Cavalry Journal* (UK) 28 (1938): 596–607.

Harmon, Ernest N. "The Second Cavalry in the Meuse-Argonne." *Cavalry Journal* (US) 31 (January 1922): 10–18.

———. "The Second Cavalry in the St. Mihiel Offensive." *Cavalry Journal* (US) 30 (July 1921): 282–89.

Harris, Moses. "With the Reserve Brigade—From Winchester to Appomattax. Fourth and Concluding Paper." *Journal of the United States Cavalry Association* 4 (March 1891): 3–26.

Hart, Basil Henry Liddell. "After Cavalry—What?" *Atlantic Monthly* 136 (September 1925): 409–418.

Hawkins, Hamilton S. "Cavalry." *Journal of the United States Cavalry Association* 26 (April 1916): 581–606.

———. "Cavalry and Mechanized Force." *Cavalry Journal* (US) 40 (September–October 1931): 19–25.

———. "German Ideas on Modern Cavalry." *Cavalry Journal* (US) 31 (April 1925): 155–68.

———. "The Importance of Modern Cavalry and Its Role as Affected by Developments in Airplane and Tank Warfare." *Cavalry Journal* (US) 35 (October 1926): 487–99.

———. "The Role of Cavalry." *Cavalry Journal* (US) 29 (October 1920): 260–65.

———. "Some Observations on the Attack by Combined Arms." *Cavalry Journal* (US) 47 (March–April 1938): 146–49.

Hayne, Paul, Jr. "Organization and Employment of Cavalry Brigades." *Journal of the United States Cavalry Association* 24 (September 1913): 202–23.

Hefferman, Leo G. "Co-operation between Cavalry and Air Service." *Cavalry Journal* (US) 34 (April 1925): 147–54.

Henderson, G. F. R., *The Science of War: A Collection of Essays and Lectures, 1891–1903.* New York: Longsmans, Green, and Co., 1908.

Hennessy, F. B. "The Aviation Squadron in the Connecticut Maneuver Campaign." *Journal of the United States Cavalry Association* 23 (November 1912): 455–477.

Herr, John K. "My Greetings to All Cavalrymen." *Cavalry Journal* (US) 47 (March–April 1938): 99.

———. "What of the Future?" *Cavalry Journal* (US) 48 (January–February 1939): 3–6.

Hickok, Howard R. "Chief of Cavalry." *Journal of the United States Cavalry Association* 18 (October 1907): 347–50.

———. "Role and Organization of Cavalry." *Journal of the United States Cavalry Association* 25 (July 1914): 50–76.

Higgs, Henry. "The Geddes Report and the Budget." *Economic Journal* 32, no. 126 (June 1922): 251–64.

Holbrook, Willard A. "A Few Words to the Cavalry." *Cavalry Journal* (US) 29 (October 1920): 249–50.

"The Horse in War." *Cavalry Journal* (US) 34 (October 1925): 412.

"Horses and Motors." Reprint from the (Springfield) *Illinois State Journal. Cavalry Journal* (US) 45 (March–April 1936): 105.

Houghton, C. F. "Attack Aviation vs. Cavalry." *Cavalry Journal* (US) 38 (April 1929): 222–27.

Hume, E. G. "Cavalry Today—Horse v. Machine?" *Cavalry Journal* (UK) 20 (1930): 180–82.

———. "Notes on Modern French Cavalry." *Cavalry Journal* (UK) 15 (January–October 1925): 27–38.

———. "Some Thoughts on Modern Reconnaissance." *Cavalry Journal* (UK) 18 (January–October 1928): 211–32.

An Infantry Transfer. "Impressions." *Cavalry Journal* (UK) 5 (July 1910): 348–53.

The Inspector of Cavalry. "What Lies Before Us." *Cavalry Journal* (UK) 1 (January 1906): 3–11.

Irvin, William R. *Combined Reconnaissance Operation by Cavalry and Aviation.* Fort Leavenworth, Kans.: Command and General Staff School, 1934.

Jacobs, Fenton. "The Cavalry." *Cavalry Journal* (US) 30 (April 1921): 124–29.

James, A. W. H. "Co-operation of Aircraft with Cavalry: Co-operation Training." *Cavalry Journal* (UK) 10 (January–October 1920): 481–87.

———. "Co-operation of Aircraft with Cavalry: General Principles." *Cavalry Journal* (UK) 10 (January–October 1920): 133–41.

———. "Co-operation of Aircraft with Cavalry: Rifles and Machine Guns *versus* Low-Flying Aircraft." *Cavalry Journal* (UK) 10 (January–October 1920): 242–53.

Johnson, Thomas J. "A Horse! A Horse! My Kingdom for a Horse." *Quartermaster Review* 16 (September–October 1936): 5–12, 77.

Joyce, Kenyon A. "The British Army Maneuvers, 1925." *Cavalry Journal* (US) 35 (January 1926): 17–28.

King, W. P. "The Fifteen Days' Training Period of the 62d Cavalry Division, Camp Meade, 1922." *Cavalry Journal* (US) 31 (January 1923): 69–73.

Lahm, Frank P. "The Relative Merits of the Dirigible Balloon." *Journal of the Military Service Institution of the United States* 48 (1911): 200–210.

Landis, J. F. Reynolds "Spirit of Sacrifice in Cavalry and Esprit de Corps in Its Officers." *Journal of the United States Cavalry Association* 20 (July 1909): 1211–19.

Lascelles, E. ff. W. "The Airship and Flying Machine in War: Their Probable Influence on the Role of Cavalry." *Cavalry Journal* (UK) 5 (April 1910): 208–12.

Leigh-Mallory, T. L. "Co-operation of Aircraft with Cavalry." *Cavalry Journal* (UK) 17 (January–October 1927): 277–83.

Lentz, Bernard. "A Justification of Cavalry." *Cavalry Journal* (US) 44 (January–February 1935): 7–11.

Lippincott, Aubrey. "The Automatic Small Arm." *Journal of the United States Cavalry Association* 13 (July 1902): 66–70.

Lodge, Henry Cabot, Jr. "Cavalry 'Marches" on Wheels: A Description of the Big Bend Portée." *Army Ordnance* 14 (November–December 1933): 135–39.

"Machines, Maneuvers, and Mud." Reprint from editorial in *Milwaukee Journal*. *Cavalry Journal* (US) 48 (January–February 1939): 73.

MacLeod, M. N. "Survey Work in Modern War." *Cavalry Journal* (UK) 11 (January–October 1921): 165–79.

———. "This Tank Business—in Fact and Fancy." *Cavalry Journal* (US) 42 (January–February 1933): 48.

Maguire, T. Miller. "The Cavalry Career of Field Marshall Viscount Allenby." *Cavalry Journal* (UK) 10 (1920): 379–84.

Mahan, Dennis Hart. *An Elementary Treatise of Advanced-Guard, Out-Post, and Detachment Service of Troops, and the Manner of Posting and Handling Them in Presence of an Enemy with a Historical Sketch of the Rise and Progress of Tactics*. New York: John Wiley, 1862.

Maude, Frederic N. *Cavalry: Its Past and Future*. London: William Clowes & Sons, Limited, 1903.

Mayhew, M. J., and G. Skeffington Smyth. "Motor Cars with the Cavalry Division." *Cavalry Journal* (UK) 4 (October 1909): 438–42.

Mayne, Charles Blair. *The Infantry Weapon and Its Use*. London: Smith, Elder, 1903.

McClure, N. F. "The Use of Cavalry." *Journal of the United States Cavalry Association* 24 (May 1914): 956–70.

McGuire, E. C. "Armored Cars in the Cavalry Maneuvers." *Cavalry Journal* (US) 39 (July 1930): 386–99.

McNair, L. J. "And Now the Autogiro." *Field Artillery Journal* 28 (1937): 3–17.

"Mechanical Warfare." *Cavalry Journal* (UK) 12 (January–October 1922): 153–56.

Merritt, Wesley. "Cavalry: Its Organization and Armament." *Journal of the Military Service Institute of the U.S.* 1 (1880): 42–52.

Mitchell, George E. "The Rout of the Turks by Allenby's Cavalry (Continued)." *Cavalry Journal* 29 (July 1920): 174–205.

Mitchell, William. *Memoirs of World War I: From Start to Finish of Our Greatest War*. New York: Random House, 1960.

———. *Our Air Force: The Keystone of National Defense*. New York: E. P. Dutton, 1921.

———. *Winged Defense: The Development and Possibilities of Modern Air Power—Economic and Military*. 1925. Reprinted, New York: Dover, 1988.

"Modern War." *Journal of the United States Cavalry Association* 20 (March 1910): 962–85.

Moffett, William A. "Some Aviation Fundamentals." *United States Naval Institute Proceedings* 51 (October 1925): 1871–81.

Morgan, George H. "Mounted Rifles." *Journal of the United States Cavalry Association* 13 (January 1903): 380–85.

———. "Some Needs of the Cavalry." *Journal of the United States Cavalry Association* 16 (October 1905): 329–31.

Moses, G. W. "Bulletin No. 18." *Journal of the United States Cavalry Association* 24 (May 1914): 908–14.

———. "Communications and Reconnaissance on the Battlefield." *Journal of the United States Cavalry Association* 23 (May 1913): 990–99.

Mulliner, A. R. "Cavalry Still an Essential Arm." *Cavalry Journal* (UK) 17 (January–October 1927): 640–47.

Nason, Leonard. "Horse and Machine." *Cavalry Journal* (US) 38 (April 1929): 192–95.

Niemann, Captain. "Airships and Cavalry in the Reconnaissance Service." *Journal of the United States Cavalry Association* 22 (March 1912): 873–77.

"Night Guard." *Cavalry Journal* (UK) 28 (1938): 170–71.

Note following "Cavalry: A Requisite." *Cavalry Journal* (US) 47 (May–June 1938): 212.

Notes on Infantry, Cavalry, and Field Artillery Lectures Delivered to Class of Provisional Second Lieutenants, Fort Leavenworth, Kansas. 1917. Washington, D.C.: Government Printing Office, 1917.

Observation Aviation March 1930. Langley Field, Va.: Air Corps Tactical School, 1930.

"An Officer Abroad." "Notes on the European War." *Journal of the United States Cavalry Association* 26 (July 1915): 53–61.

An Officer of High Rank. "What Has the World's War Taught Us Up to the Present Time That Is New in a Military Way." *Journal of the United States Cavalry Association* 26 (July 1915): 97–99.

One of Our Cavalry Officers. "Service with a French Cavalry Regiment." *Journal of the United States Cavalry Association* 25 (October 1914): 226–34.

One of the Faithful. [pseud.] "Faith in and a Doctrine for the Cavalry Service." *Cavalry Journal* (US) 36 (April 1927): 227–31.

Osborne, Rex. "Operations of the Mounted Troops of the Egyptian Expeditionary Force (continued)." *Cavalry Journal* (UK) 13 (January–October 1923): 21–41.

Other Arms Air Corps. Fort Riley, Kans.: Academic Division, The Cavalry School, 1931–32.

Parker, James. "The Cavalryman and the Rifle." *Cavalry Journal* (US) 37 (July 1928): 362–67.

———. "The Retention of the Saber as a Cavalry Weapon." *Journal of the United States Cavalry Association* 14 (October 1903): 354–58.

———. "Saber Versus Revolver Versus Carbine." *Journal of the United States Cavalry Association* 17 (July 1906): 35–41.

———. "The Value of Cavalry as Part of Our Army." *Journal of the United States Cavalry Association* 25 (July 1914): 5–13.

Patton, George S., Jr. "Motorization and Mechanization in the Cavalry." *Cavalry Journal* (US) 39 (July 1930): 331–348.

———. "The 1929 Cavalry Division Maneuvers." *Cavalry Journal* (US) 39 (January 1929): 7–15.

———. "What the World War Did for Cavalry." *Cavalry Journal* (US) 31 (April 1922): 165–72.

Patton, George S., Jr., and C. C. Benson. "Mechanization and Cavalry." *Cavalry Journal* (US) 39 (April 1930): 234–40.

Perry, Redding F. "The 2d Cavalry in France." *Cavalry Journal* (US) 37 (January 1928): 27–41.

Pershing, John J. "A Message to the Cavalry." *Cavalry Journal* (US) 29 (April 1920): 5–6.

Pierce, John T. *Place of a Mechanized Force in Our Cavalry Organization.* Fort Leavenworth, Kans.: Command and General Staff School, 1932.

"Preface." *Cavalry Journal* (UK) 6 (January 1911): B–B2.

Preston, R. M. *The Desert Mounted Corps: An Account of the Cavalry Operations in Palestine and Syria, 1917–1918.* London: Constable, 1921.

"Progress and Discussion." *Cavalry Journal* (US) 39 (January 1930): 118–25.

Prosser, Walter E. "A Discussion of the War Balloon and Similar Craft, and the Best Methods of Attack by Artillery." *Journal of the United States Artillery* 34 (July–August 1910): 257–62.

"Reconnaissance." *Cavalry Journal* (US) 30 (October 1921): 353.

Reeves, James H. "Cavalry Raids." *Journal of the United States Cavalry Association* 10 (September 1897): 232–47.

Regulations and Programm of Instruction of the U.S. Infantry and Cavalry School. Fort Leavenworth, Kans.: United States Infantry and Cavalry School, 1895.

Reilly, Henry J. "Cavalry in Modern War." *Journal of the United States Cavalry Association* 27 (November 1916): 294–97.

———. "Cavalry in the Great War." *Journal of the United States Cavalry Association* 27 (April 1917): 477–82.

Rhodes, Charles D. "The Cavalry of Today." *Journal of the United States Cavalry Association* 24 (November 1913): 359–71.

———. "The Duties of Cavalry in Modern Wars." *Journal of the United States Cavalry Association* 6 (June 1893): 172–81.

Riedl, Nickolaus. "Cavalry in War." *Journal of the United States Cavalry Association* 23 (September 1912): 290–307.

Roscoe, Daniel L. "The Effect of Aeroplanes upon Cavalry Tactics." *Journal of the United States Cavalry Association* 24 (March 1914): 856–58.

Rubottom, Holland. "Cavalry Reconnaissance and Transmission of Information by Modern Methods." *Journal of the United States Cavalry Association* 23 (July 1912): 25–29.

Russell, George M. "Intelligence for Cavalry." *Cavalry Journal* (US) 29 (October 1920): 254–59.

Ryan, W. Michael. "The Invasion Controversy of 1906–1908: Lieutenant-Colonel Charles a Court Repington and British Perceptions of the German Menace." *Military Affairs* 44, no. 1 (February 1980): 8–12.

Sanden, V. "Cavalry and Aircraft in the Service of Reconnaissance." *Journal of the United States Cavalry Association* 23 (March 1913): 826–32.

Sayre, Farrand. "Cavalry Drill and Organization." *Journal of the United States Cavalry Association* 26 (July 1915): 5–17.

Schwarzkoff, Olaf. "The Changed Status of the Horse in War." *Journal of the United States Cavalry Association* 26 (January 1916): 335–53.

Scott, Charles L. "Are More Changes Need in Our Horsed Cavalry Regiment Now?" *Cavalry Journal* (US) 44 (September–October 1935): 42.

Sherman, William C. "Cavalry and Aircraft." *Cavalry Journal* (US) 30 (January 1921): 26–30.

Sherman, William Carr. *Air Warfare.* New York: Ronald, 1926.

Sherrill, Stephen H. "The Experiences of the First American Troop of Cavalry to Get into Action in the World War." *Cavalry Journal* (US) 32 (April 1923): 153–59.

"A Study of Patrol Work." *Cavalry Journal* (UK) 8 (January–October 1913): 420–27.

Smith, John B. "The Effect of Mechanization Upon Cavalry." *Cavalry Journal* (US) 40 (November–December 1931): 21–24.

Stacey, Cyril. "The Passing of the Cavalry, 1939." *Cavalry Journal* (UK) 29 (January–October 1939): 566.

Stackpole, Edward J., Jr. "The National Guard Cavalry." *Cavalry Journal* (US) 47 (March–April 1938): 100–103.

Stockhausen, V. "Airships in War." *Cavalry Journal* (US) 20 (November 1909): 575–84.

Strettell, C. B. Dashwood. "Cavalry in Open Warfare, Illustrated by the Operations Leading Up to the Occupation of Mosul in November, 1918." *Journal of the Royal United Service Institution* 66 (1921): 598–617.

Sumner, E. V. "American Practice and Foreign Theory." *Journal of the United States Cavalry Association* 3 (June 1890): 142–50.

Tactics and Technique of Cavalry. Harrisburg, Pa.: Telegraph Press, 1937.

Tactics and Technique of Cavalry: A Text and Reference Book of Cavalry Training (Includes All Technical Changes to June, 1937). 8th ed. Harrisburg, Pa.: Military Service Publishing, 1937.

Taylor, John R. M. "Cavalry and the Aeroplane." *Journal of the United States Infantry Association* 6 (July 1909): 84–88.

Taylor, William J. "The 305th Cavalry Command Post Exercise." *Cavalry Journal* (US) 44 (May–June 1935): 24–25.

Tilney, W. A. "Aerial Reconnaissance in War." *Cavalry Journal* (UK) 6 (January–October 1911): 12–13.

Tittinger, A. J. "The Future of Cavalry." *Cavalry Journal* (US) 29 (April 1920): 67–69.

"Topics of the Day: Air Service and Other Auxiliaries for Cavalry." *Cavalry Journal* (US) 31 (October 1922): 412.

"Topics of the Day: Cavalry Discussed by Royal United Service Institution." *Cavalry Journal* (US) 31 (April 1922): 201–2.

"Topics of the Day: The Expedition into Mexico to Recover the Pierson Airplane." *Cavalry Journal* (US) 30 (July 1921): 304–7.

"Topics of the Day: Extracts from the Annual Report of the Chief of Cavalry, Major General Herbert B. Crosby." *Cavalry Journal* (US) 37 (January 1928): 108–9.

"Topics of the Day: Reorganizing of the British Cavalry." *Cavalry Journal* (US) 36 (July 1927): 479–81.

"Topics of the Day: Two More Types of Guns for Cavalry." *Cavalry Journal* (US) 37 (July 1928): 436–37.

United States Air Corps Tactical School. *Observation Aviation March 1930.* Langley Field, Va.: Air Corps Tactical School, 1930.

U.S. Army Command and General Staff School. *Reconnaissance Security Marches Halts (Tentative).* Fort Leavenworth, Kans.: Command and General Staff School Press, 1937.

———. *Tables of Organization.* Fort Leavenworth, Kans.: Command and General Staff School Press, 1937.

————. *Tactical Employment of Cavalry.* Fort Leavenworth, Kans.: Command and General Staff School Press, 1935.

————. *Tactical Principles and Decisions: Volume 1 Marches, Halts, and Reconnaissance and Security, 1920.* Rev. ed. Fort Leavenworth, Kans.: General Service Schools Press, 1922.

U.S. War Department. *Field Service Regulations, United States Army, 1914.* Washington, D.C.: Government Printing Office, 1914.

————. *Field Service Regulations, United States Army, 1923.* Washington, D.C.: Government Printing Office, 1924.

"Value of Animals in Modern Warfare." *Cavalry Journal* (US) 47 (March–April 1938): 109.

Vedette. "Aeroplanes and Their Influence upon Cavalry Training." *United Service Magazine* 83, no. 1005 (1912): 555–57.

Velox [pseud.]. "A Chief of Cavalry." *Journal of the United States Cavalry Association* 15 (April 1905): 944–46.

Vidmer, George. "The Service Pistol and Its Caliber." *Journal of the United States Cavalry Association* 16 (October 1905): 181–88.

Vindex. "Modern Inventions and the Functions of Cavalry." *Cavalry Journal* (UK) 7 (January–October 1912): 348–52.

Von Ammon. "Cavalry Lessons of the Great War from German Sources." *Cavalry Journal* (US) 30 (October 1921): 358–64.

Wade, Jack. "Lest Ye Forget." *Cavalry Journal* (US) 31 (July 1922): 312.

Wagner, Arthur L. *A Catechism of Outpost Duty, Including Advance Guards, Rear Guards, and Reconnaissance.* Kansas City, Mo.: Hudson-Kimberly, 1895.

————. *Organization and Tactics.* New York: B. Westermann, 1895.

————. *The Service of Security and Information.* Kansas City, Mo.: Hudson-Kimberly, 1903.

Waldron, F. E. "Aeroplanes and Cavalry." *Cavalry Journal* (UK) 8 (January–October 1913): 313–19.

Walker, Kirby. "Cavalry in the World War." *Cavalry Journal* (US) 33 (January 1924): 11–23.

"War and the 'Arme Blanche.'" *Cavalry Journal* (UK) 5 (July 1910): 283–87.

War Department: Office of the Chief of Staff. *Cavalry Drill Regulations United States Army 1916, Corrected to December 31, 1917 (Changes nos. 1 and 2).* Washington, D.C.: Government Printing Office, 1918.

War Office. *Cavalry Training, Volume I: Training, 1931.* London: His Majesty's Stationery Office, 1931.

————. *Cavalry Training, Volume II: War, 1929.* London: His Majesty's Stationery Office, 1929.

————. *Field Service Regulations Volume I: Organization and Administration, 1930.* London: His Majesty's Stationery Office, 1930.

————. *Field Service Regulations Volume II: Operations, 1924.* London: His Majesty's Stationery Office, 1924.

————. *Field Service Regulations Volume II: Operations—General, 1935.* London: His Majesty's Stationery Office, 1935.

————. *Field Service Regulations Volume II: Operations, 1929.* London: His Majesty's Stationery Office, 1929.

———. *Field Service Regulations Volume III: Operations—Higher Formations, 1935.* London: His Majesty's Stationery Office, 1935.

Weir, G. A. "Some Reflections on the Cavalry Campaign in Palestine." *Journal of the Royal United Service Institution* 67 (1922): 219–35.

Whiting, E. M. "Protection from Enemy Aircraft." *Cavalry Journal* (US) 37 (January 1928): 103–6.

Wisser, John P. "The Tactical and Strategical use of Dirigible Balloons and Aeroplanes." *Journal of the United States Cavalry Association* 21 (November 1910): 412–24.

Wood, Leonard. "Cavalry's Role in the Reorganization." *Cavalry Journal* (US) 29 (July 1920): 113–15.

X [pseud]. "Chief of Cavalry: Shall We Have a Chief of Cavalry?" *Journal of the United States Cavalry Association* 17 (January 1907): 556–58.

NEWSPAPERS AND MAGAZINES

Aero and Hydro America's Aviation Weekly
American Aeronaut
Boston Daily Globe
Chicago Daily Tribune
Christian Science Monitor
Detroit Free Press
Hartford Courant
Independent . . . Devoted to the Consideration of Politics, Social and Economic Tendencies, History, Literature, and the Arts
Indianapolis Star
Kansas City Star
Literary Digest
Living Age
Los Angeles Times
Manchester Guardian
Nashville Tennessean and Nashville American
Nation and the Athenaeum
Newsweek
New York Times
New York Tribune
Observer
Popular Science Monthly
San Francisco Chronicle
St. Louis Post-Dispatch
Times (London)
Town and Country
Wall Street Journal
Washington Post

SECONDARY SOURCES

Anglesey, Marquess of. *A History of British Cavalry.* 8 vols. London: Secker and War-burg, 1986.

Badsey, Stephen. "The Boer War (1899–1902) and British Cavalry Doctrine: A Re-evalu-ation." *Journal of Military History* 71 (January 2007): 75–97.

———. *Doctrine and Reform in the British Cavalry, 1880–1918.* Aldershot, England: Ash-gate, 2008.

Ball, Durwood. *Army Regulars on the Western Frontier, 1848–1861.* Norman: University of Oklahoma Press, 2001.

Barr, Ronald J. *The Progressive Army: U.S. Army Command and Administration, 1870–1914.* New York: St. Martin's Press, 1998.

Bauer, Martin. *Resistance to New Technology: Nuclear Power, Information Technology, and Biotechnology.* Cambridge: Cambridge University Press, 1995.

Biddle, Stephen. *Afghanistan and the Future of Warfare: Implications for Army and De-fense Policy.* Carlisle, Pa.: Strategic Studies Institute, 2002.

Bielakowski, Alexander M. "United States Army Cavalry Officers and the Issue of Mech-anization, 1920 to 1942." PhD diss., Kansas State University, 2002.

———. *U.S. Cavalryman, 1891–1920.* Oxford: Osprey, 2004.

Bijker, Wiebe, Thomas P. Hughes, and Trevor Pinch. *The Social Construction of Tech-nological Systems: New Directions in the Sociology and History of Technology.* Cam-bridge, Mass.: MIT Press, 1997.

Bijker, Wiebe E., and John Law. *Shaping Technology/Building Society: Studies in Socio-technical Change.* Cambridge, Mass.: MIT Press, 1992.

Bond, Brian. "Doctrine and Training in the British Cavalry, 1870–1914." In *The Theory and Practice of War,* edited by Michael Howard, 95–125. London: Cassell, 1965.

Brereton, T. R. *Educating the U.S. Army: Arthur Wagner and Reform, 1875–1905.* Lincoln: University of Nebraska Press, 2000.

Brett, R. Dallas. *History of British Aviation, 1908–1914.* 2 vols. London: Aviation Book Club, 1934.

Brooks, Peter W. *Cierva Autogiros: The Development of Rotary-Wing Flight.* Washing-ton, D.C.: Smithsonian Institution Press, 1988.

Budiansky, Stephen. *Air Power: The Men, Machines, and Ideas that Revolutionized War, from Kitty Hawk to Gulf War II.* New York: Viking, 2004.

Cameron, Robert Stewart. "Americanizing the Tank: U.S. Army Administration and Mechanized Development within the Army, 1917–1943." PhD diss., Temple Univer-sity, 1994.

Chandler, Charles de Forest and Frank Purdy Lahm. *How Our Army Grew Wings: Air-men and Aircraft before 1914.* New York: Ronald Press, 1943.

Charnov, Bruce H. *From Autogiro to Gyroplane: The Amazing Survival of an Aviation Technology.* Westport, Conn.: Praeger, 2003.

Connor, Roger Douglas. "Grasshoppers and Jump-Takeoff: The Autogiro Programs of the U.S. Army Air Corps." Paper presented at the Sixty-Second Annual Forum and Technology Display of the American Helicopter Society, Phoenix, Arizona, May 9–11, 2006.

Cooke, James J. *Billy Mitchell.* Boulder, Colo.: Lynne Rienner, 2002.

Coopersmith, Jonathan. "Failure & Technology." *Japan Journal for Science, Technology & Society* 18 (2009): 93–118.

Corn, Joseph J., ed. *Imagining Tomorrow: History, Technology, and the American Future.* Cambridge, Mass.: MIT Press, 1986.

———. *The Winged Gospel: America's Romance with Aviation, 1900–1950.* New York: Oxford University Press, 1983.

Corum, James S., and Wray R. Johnson. *Airpower in Small Wars: Fighting Insurgents and Terrorists.* Lawrence: University Press of Kansas, 2003.

Cowan, Ruth Schwartz. "The Consumption Junction: A Proposal for Research Strategies in the Sociology of Technology." In *The Social Construction of Technological Systems,* edited by Wiebe E. Bijker, Thomas P. Hughes, and Trevor Pinch, 261–280. Cambridge, Mass.: MIT Press, 1989.

Crosby, Francis. *A Handbook of Fighter Aircraft.* London: Hermes House, 2002.

Dimarco, Louis A. *War Horse: A History of the Military Horse and Rider.* Yardley, Pa.: Westholme, 2008.

Driver, Hugh. *Birth of Military Aviation: Britain, 1903–1914.* Woodbridge, UK: The Royal Historical Society: Boydell Press, 1997.

Edgerton, David. *England and the Aeroplane: An Essay on a Militant and Technological Nation.* Basingstoke, UK: Macmillan in Association with the Centre for the History of Science, Technology and Medicine, University of Manchester, 1991.

———. *The Shock of the Old: Technology and Global History since 1900.* Oxford: Oxford University Press, 2007.

Farrell, Theo, and Terry Terriff. *The Sources of Military Change: Culture, Politics, Technology.* Boulder, Colo.: Lynne Rienner, 2002.

Finney, Robert T. *History of the Air Corps Tactical School, 1920–1940.* Washington, D.C.: Air Force History and Museums Program, 1998.

Fishbein, Samuel B. *Edward Steichen: A Brief Biographical Sketch of This Pioneer and the History of Photographic Technology Prior to and During World War I.* National Air and Space Museum Smithsonian Institution, 1994.

Flick, Carlos. "The Movement for Smoke Abatement in 19th-Century England." *Technology and Culture* 21 (January 1980): 29–50.

Forester, C. S. *The General.* 1936. Reprint, Annapolis, Md.: Nautical and Aviation Publishing, 1982.

Friedel, Robert. *Zipper: An Exploration in Novelty.* New York: W. W. Norton, 1994.

Fries, Russell I. "British Response to the American System: The Case of the Small-Arms Industry After 1869." *Technology and Culture* 16 (July 1975): 377–403.

Gillie, Mildred Hanson. *Forging the Thunderbolt: A History of the Development of the Armored Force.* Harrisburg, Pa.: Military Service Publishing, 1947.

Goldstein, Carolyn M. "From Service to Sales: Home Economics in Light and Power, 1920–1940." *Technology and Culture* 38 (1997): 121–52.

Gollin, Alfred. "England Is No Longer an Island: The Phantom Airship Scare of 1909." *Albion: A Quarterly Journal Concerned with British Studies* 13, no. 1 (Spring 1981): 43–57.

———. *The Impact of Air Power on the British People and Their Government, 1909–14.* Stanford, Calif.: Stanford University Press, 1989.

———. "The Mystery of Lord Haldane and Early British Military Aviation." *Albion: A Quarterly Journal Concerned with British Studies* 11, no. 1 (Spring 1979): 46–65.

———. *No Longer an Island: Britain and the Wright Brothers, 1902–1909.* Stanford, Calif.: Stanford University Press, 1984.

Greer, Thomas H. *The Development of Air Doctrine in the Army Air Arm, 1917–1941.* Washington, D.C.: Office of Air Force History, 1985.

Gregory, H. F. *Anything a Horse Can Do: The Story of the Helicopter.* New York: Reynal and Hitchcock, 1944.

Hampton, Mark. "Inventing David Low: Self-Presentation, Caricature and the Culture of Journalism in Mid-Twentieth Century Britain." *Twentieth Century British History* 20, no. 4 (2009): 482–512.

Hart, Basil H. Liddell. *The Tanks: The History of the Royal Tank Regiment and Its Predecessors, Heavy Branch Machine-Gun Corps, Tank Corps, and Royal Tank Corps 1914-1945.* Vol 1. London: Cassell, 1959.

Herr, John K., and Edward S. Wallace. *The Story of the U.S. Cavalry, 1775–1942.* Boston: Little, Brown, 1953.

Higham, Robin. *100 Years of Air Power and Aviation.* College Station: Texas A&M University Press, 2003.

Hinkle, Stacy. "Wings and Saddles: The Air and Cavalry Punitive Expedition of 1919." *Southwestern Studies* 5 (1967).

———. *Wings over the Border: The Army Air Service Armed Patrol of the United States-Mexico Border, 1919–1921.* El Paso: Western Press, University of Texas, 1970.

Hinton, Harold. *Air Victory: The Men and the Machines.* New York: Harper and Brothers, 1948.

Hofmann, George F. *Through Mobility We Conquer: The Mechanization of U.S. Cavalry.* Lexington: University Press of Kentucky, 2006.

Holley, I. B. *Army Air Forces Historical Studies: No. 44—Evolution of the Liaison-Type Airplane, 1917–1944.* Washington, D.C.: AAF Historical Office, 1946.

Holley, I. B., Jr. *Ideas and Weapons: Exploitation of the Aerial Weapon by the United States during World War I; A Study in the Relationship of Technological Advance, Military Doctrine, and the Development of Weapons.* New Haven, Conn.: Yale University, 1953.

Howard, Michael. *The Theory and Practice of War.* Bloomington: Indiana University Press, 1965.

Hugill, Peter J. *Global Communications since 1844: Geopolitics and Technology.* Baltimore, Md.: Johns Hopkins University Press, 1999.

Huntington, Samuel. *The Soldier and the State: The Theory and Politics of Civil-Military Relations.* Cambridge, Mass.: Belknap Press of Harvard University Press, 1957.

Hurley, Alfred F. *Billy Mitchell: Crusader for Air Power.* New York: Franklin Watts, 1964.

James, Lawrence. *Imperial Rearguard: Wars of Empire, 1919–1985.* London: Brassey's Defence, 1988.

Janowitz, Morris. *The Professional Soldier: A Social and Political Portrait.* Glencoe, Ill.: Free Press, 1960.

Jarymowycz, Roman. *Cavalry from Hoof to Track.* Westport, Conn.: Praeger, 2008.

Jeffery, Keith. *The British Army and the Crisis of Empire, 1918–22.* Manchester, UK: Manchester University Press, 1984.

Johnson, David E. *Fast Tanks and Heavy Bombers: Innovation in the U.S. Army, 1917–1945.* Ithaca, N.Y.: Cornell University Press, 1998.

Johnson, Herbert A. *Wingless Eagle: U.S. Army Aviation through World War I.* Chapel Hill: University of North Carolina Press, 2001.

Katzenbach, Edward. "The Horse Cavalry in the Twentieth Century: A Study in Policy Response." *Public Policy* 8 (1958): 120–49.

Kennett, Lee B. *The First Air War, 1914–1918.* New York: Free Press, 1991.

Kenyon, David. *Horsemen in No Man's Land: British Cavalry and Trench Warfare, 1914–1918.* Barnsley, South Yorkshire, UK: Pen and Sword, 2011.

Kirwan, Elizabeth T. "The Cavalry School Library, Fort Riley, Kansas." *Library Journal* 62 (March 1937): 196–98.

Kline, Ronald R. *Consumers in the Country: Technology and Social Change in Rural America.* Baltimore, Md.: Johns Hopkins University Press, 2000.

Lawson, Eric, and Jane Lawson. *The First Air Campaign: August 1914–November 1919.* Conshohocken, Pa.: Combined Books, 1996.

Leslie, Stuart W. "Charles F. Kettering and the Copper-Cooled Engine." *Technology and Culture* 20 (October 1979): 752–76.

Lewis, Peter. *The British Bomber Since 1914: Fifty Years of Design and Development.* London: Putnam, 1967.

Lewis, W. David. "The Autogiro Flies the Mail! Eddie Rickenbacker, Johnny Miller, Eastern Air Lines, and Experimental Airmail Service with Rotorcraft, 1939–1940." In *Biographical Essays in Honor of the Centennial of Flight, 1903–2003*, edited by Virginia Parker Dawson and Mark D. Bowles, 69–86. Washington, D.C.: National Aeronautics and Space Administration History Division, Office of External Relations, 2005.

Linn, Brian McAllister. *The Echo of Battle: The Army's Way of War.* Cambridge, Mass.: Harvard University Press, 2007.

Mackworth-Praed, Ben, ed. *Aviation: The Pioneer Years.* Secaucus, N.J.: Chartwell Books, 1990.

Maree, D.R. "Bicycles in the Anglo-Boer War of 1899–1902." *Military History Journal* 4 (June 1977), 1–13.

Mead, Peter. *The Eye in the Air: History of Air Observation and Reconnaissance for the Army, 1785–1945.* London: Her Majesty's Stationery Office, 1983.

Melhorn, Charles M. *Two-Block Fox: The Rise of the Aircraft Carrier, 1911–1929.* Annapolis, Md.: Naval Institute Press, 1974.

Mets, David R. *Airpower and Technology: Smart and Unmanned Weapons.* Westport, Conn.: Praeger, 2009.

Millett, Allan R., and Peter Maslowski. *For the Common Defense: A Military History of the United States of America.* Rev. ed. New York: Free Press, 1994.

Millis, Walter. *Arms and Men: A Study in American Military History.* 1956. Reprint, New Brunswick, N.J.: Rutgers University Press, 1984.

Montross, Lynn. *Cavalry of the Sky: The Story of U.S. Marine Combat Helicopters.* New York: Harper and Brothers, 1954.

Morrow, John H., Jr. *The Great War in the Air: Military Aviation from 1909 to 1921.* Washington D.C.: Smithsonian Institute Press, 1993.

Muller, Richard R. "Close Air Support: The German, British and American Experiences, 1918–1941." In *Military Innovation in the Interwar Period,* edited by Williamson Murray and Allan R. Millett, 144–90. Cambridge: Cambridge University Press, 1996.

Munson, Kenneth. *Bombers: Patrol and Reconnaissance Aircraft.* London: Bounty Books, 2012.

———. *Fighters: Attack and Training Aircraft, 1914–19.* London: Bounty Books, 2004.

Murphy, Justin D., and Matthew A. McNiece. *Military Aircraft 1919–1945: An Illustrated History of Their Impact.* Santa Barbara, Calif.: ABC-Clio, 2009.

Murray, Williamson, and Allan R. Millett, eds. *Military Innovation in the Interwar Period.* Cambridge: Cambridge University Press, 1996.

Nedialkov, Dimitar. *Genesis of Airpower.* Sofia: Pensoft, 2004.

Nenninger, Timothy K. "American Military Effectiveness in the First World War." In *Military Effectiveness,* Vol. 1, *The First World War,* edited by Allan R. Millett and Williamson Murray, 116–56. Boston: Unwin Hyman, 1988.

———. *The Leavenworth Schools and the Old Army: Education, Professionalism, and the Officer Corps of the United States Army, 1881–1918.* Westport, Conn.: Greenwood Press, 1978.

Neufeld, Michael J. *The Rocket and the Reich: Peenmunde and the Coming of the Ballistic Missile Era.* New York: Free Press, 1995.

Newark, Peter. *Sabre and Lance: An Illustrated History of Cavalry.* Poole, Dorset, UK: Blandford Press, 1987.

Odom, William. *After the Trenches: The Transformation of U.S. Army Doctrine, 1918–1939.* College Station: Texas A&M University Press, 1999.

Omissi, David E. *Air Power and Colonial Control: The Royal Air Force, 1919–1939.* Manchester, UK: Manchester University Press, 1990.

Oudshoorn, Nelly, and Trevor Pinch. *How Users Matter: The Co-Construction of Users and Technologies.* Cambridge, Mass.: MIT Press, 2003.

Paris, Michael. *Winged Warfare: The Literature and Theory of Aerial Warfare in Britain, 1859–1917.* Manchester: University of Manchester, 1992.

Phillips, Gervase. "Scapegoat Arm: Twentieth-Century Cavalry in Anglophone Historiography." *Journal of Military History* 71 (January 2007): 37–74.

Perkins, John H. "Reshaping Technology in Wartime: The Effect of Military Goals on Entomological Research and Insect-Control Practices." *Technology and Culture* 19 (April 1978): 169–86.

Post, Robert C. "The Page Locomotive: Federal Sponsorship of Invention in Mid-19th-Century America." *Technology and Culture* 13 (April 1972): 140–69.

Ragsdale, Kenneth Baxter. *Wings over the Mexican Border: Pioneer Aviation in the Big Bend.* Austin: University of Texas Press, 1984.

Randall, Adrian. "Reinterpreting 'Luddism': Resistance to New Technology in the British Industrial Revolution." In *Resistance to New Technology: Nuclear Power, Information Technology, and Biotechnology,* edited by Martin Bauer, 57–79. Cambridge: Cambridge University Press, 1995.

Reardon, Carol. *Soldiers and Scholars: The U.S. Army and the Uses of Military History, 1865–1920.* Lawrence: University Press of Kansas, 1990.

Reid, Brian Holden. "'A Signpost That Was Missed'?: Reconsidering British Lessons from the American Civil War." *Journal of Military History* 70 (April 2006): 385–414.

Rittgers, Sarah Janelle. "From Galloping Hooves to Rumbling Engines: Organizational Responses to Technology in the U.S. Horse Cavalry." PhD diss., George Washington University, 2003.

Robson, Brian. *Crisis on the Frontier: The Third Afghan War and the Campaign in Waziristan, 1919–20.* Staplehurst: Spellmount, 2007.

Roe, Andrew. *Waging War in Waziristan: The British Struggle in the Land of Bin Laden, 1849–1947.* Lawrence: University Press of Kansas, 2010.

Rowe, David. *Head Dress of the British Heavy Cavalry: Dragoon Guards, Household and Yeomanry Cavalry 1842–1934.* Atglen, Pa.: Schiffer Military History, 1999.

Schatzberg, Eric. *Wings of Wood, Wings of Metal: Culture and Technical Choice in American Airplane Materials, 1914–1945.* Princeton, N.J.: Princeton University Press, 1999.

Seymour-Ure, Colin, and Jim Schoff. *David Low.* London: Secker and Warburg, 1985.

Skelton, William B. "Professionalization in the U.S. Army Officer Corps during the Age of Jackson." *Armed Forces & Society* 1 (1975): 443–71.

———. "Samuel P. Huntington and the Roots of the American Military Tradition." *Journal of Military History* 60 (April 1996): 325–38.

Smith, Malcolm. *British Air Strategy between the Wars.* Oxford: Clarendon Press, 1984.

Stanton, Doug. *Horse Soldiers: The Extraordinary Story of a Band of U.S. Soldiers Who Rode to Victory in Afghanistan.* New York: Scribner, 2009.

Staudenmaier, John M. *Technology's Storytellers: Reweaving the Human Fabric.* Cambridge, Mass.: MIT Press, 1985.

Stokesbury, James L. *A Short History of Airpower.* London: Robert Hale, 1986.

Stubbs, Mary Lee, and Stanley Russell Connor. *Armor-Cavalry Part I: Regular Army and Army Reserve.* Washington, D.C.: Office of the Chief of Military History, United States Army, 1969.

Swanborough, Gordon, and Peter M. Bowers. *United States Military Aircraft since 1909.* Washington, D.C.: Smithsonian Institution Press, 1989.

Taylor, Michael. *Jane's Encyclopedia of Aviation.* New York: Portland House, 1989.

———. *Jane's Fighting Aircraft of World War I.* 1919. Reprint, London: Random House, 2001.

Tedesco, Vincent J., III. "'Greasy Automatons' and the 'Horsey Set': The U.S. Cavalry and Mechanization, 1928–1940." Master's thesis, Pennsylvania State University, 1995.

Templewood, Samuel John Gurney Hoare. *Empire of the Air: The Advent of the Air Age, 1922–1929.* London: Collins, 1957.

Thayer, Lucien H. *America's First Eagles: The Official History of the U.S. Air Service, A.E.F. (1917-1918).* San Jose, Calif. and Mesa, Ariz.: Bender and Champlin Fighter Aces Museum Press, 1983.

Tobin, James. *To Conquer the Air: The Wright Brothers and the Great Race for Flight.* New York: Free Press, 2003.

Towle, Philip Anthony. *Pilots and Rebels: The Use of Aircraft in Unconventional Warfare, 1918–1988.* London: Brassy's, 1989.

Turner Publishing Co. *11th U.S. Cavalry, Blackhorse.* Paducah, Ky.: Turner, 1990.

Utley, Robert M. *Frontier Regulars: The United States Army and the Indian, 1866–1891.* New York: Macmillan, 1973.

———. *Frontiersmen in Blue: The United States Army and the Indian, 1848–1865.* New York: Macmillian, 1967.

Urwin, Gregory J. W. *The United States Cavalry: An Illustrated History.* Pode, UK: Blandford, 1983.

Warner, Philip. *The British Cavalry.* London: J. M. Dent, 1984.

Weigley, Russell F. *History of the United States Army.* New York: Macmillan, 1967.

White, Elwood L. *Air Power and Warfare: A Supplement.* Colorado Springs, Colo.: United States Air Force Academy, 2002.

Winchester, Jim. *American Military Aircraft: A Century of Innovation.* New York: Barnes and Noble Books, 2005.

Winton, Harold. *To Change an Army: General Sir John Burnett-Stuart and British Armored Doctrine, 1927–1938.* Lawrence: University of Kansas Press, 1988.

Wohl, Robert. *A Passion for Wings: Aviation and the Western Imagination, 1908–1918.* New Haven, Conn.: Yale University Press, 1994.

———. *The Spectacle of Flight: Aviation and the Western Imagination, 1920–1950.* New Haven, Conn.: Yale University Press, 2005.

Woodward, David R. *Hell in the Holy Land: World War I in the Middle East.* Lexington: University Press of Kentucky, 2013.

Wormser, Richard. *The Yellowlegs: The Story of the United States Cavalry.* Garden City, N.Y.: Doubleday, 1966.

Wrangel, Alexis. *The End of Chivalry: The Last Great Cavalry Battles, 1914–1918.* New York: Hippocrene Books, 1982.

INDEX

CPSIA information can be obtained
at www.ICGtesting.com
Printed in the USA
LVHW112317140219
607645LV00004B/11/P